T0214030

Mechanical Engineering Series

Series Editor
Francis A. Kulacki, Department of Mechanical Engineering,
University of Minnesota, Minneapolis, MN, USA

The Mechanical Engineering Series presents advanced level treatment of topics on the cutting edge of mechanical engineering. Designed for use by students, researchers and practicing engineers, the series presents modern developments in mechanical engineering and its innovative applications in applied mechanics, bioengineering, dynamic systems and control, energy, energy conversion and energy systems, fluid mechanics and fluid machinery, heat and mass transfer, manufacturing science and technology, mechanical design, mechanics of materials, micro- and nano-science technology, thermal physics, tribology, and vibration and acoustics. The series features graduate-level texts, professional books, and research monographs in key engineering science concentrations.

More information about this series at http://www.springer.com/series/1161

Zainul Huda

Mechanical Behavior
of Materials

Fundamentals, Analysis, and Calculations

 Springer

Zainul Huda
King Abdulaziz University
Jeddah, Saudi Arabia

ISSN 0941-5122   ISSN 2192-063X   (electronic)
Mechanical Engineering Series
ISBN 978-3-030-84929-0   ISBN 978-3-030-84927-6   (eBook)
https://doi.org/10.1007/978-3-030-84927-6

This Springer imprint is published by the registered company Springer Nature Switzerland AG
The registered company address is: Gewerbestrasse 11, 6330 Cham, Switzerland

Preface

Mechanical behavior of materials is taught as a core course in undergraduate materials/mechanical engineering programs in almost all reputed universities. This book, therefore, presents fundamentals and quantitative analysis of mechanical behavior of materials covering engineering mechanics and materials, deformation behavior, fracture mechanics, and failure design.

This textbook provides a holistic understanding of mechanical behavior of materials and enables critical thinking through mathematical modeling and problem solving. The salient features of this text include: 200 calculations/worked examples, 120 diagrams/photographs, and over 260 equations/formulae on mechanics and materials, including fatigue, creep, and the mechanical behavior of composite materials. The book is equally useful to both engineering-degree students and production/design engineers practicing in manufacturing industry; the latter may specially benefit from highly practical stuff on shaft design and failure design.

The book is divided into three parts. Part 1 (Chaps. 1—5) introduces readers to material failures, crystal structure and defects, dislocation motion and deformation, strengthening mechanisms and mechanical testing of materials, and metallic and non-metallic materials. In this part, a detailed analysis of deformation and strengthening mechanisms is included. Part 2 (Chaps. 6–10) deals with stresses, strains, elasticity, plasticity, and stresses/torsion in static shafts. Here, special emphases are given to stress-strain relations, principal and complex stresses, and elastic and plastic deformations, including their engineering analyses and problem solving.

Finally, Part 3 (Chaps. 11–15) deals with failure theories and design, fracture mechanics, fatigue and creep, and the mechanical behavior of composite materials. Here, the failure theories have been applied to design shafts, and fracture mechanics design philosophy are applied to design engineering components, including pressure vessels. In this part, the mechanical behavior of composite materials is also quantitatively discussed and analyzed so as to enable the reader to solve numerical problems.

This book contains 15 chapters. Each chapter first introduces readers to the technological importance of the topic and provides basic concepts with diagrammatic illustrations, and then its engineering analysis/mathematical modeling along with calculations are presented. An attempt is made to use SI units in each mathematical model throughout the text. There are altogether **200 worked examples/calculations, 120 diagrams/photographs,** and over **260 mathematical models/equations.** In particular, the worked examples cover solutions to both simple and difficult-to-solve problems. In order to best benefit from this book, the student/ reader is advised to first prepare a list of relevant formulae and then attempt to solve a problem (worked example) while hiding its solution; they must try to solve the problem on their own by analyzing the data. In case the reader is stuck during solving the problem, s/he may refer to its solution. *Question and Exercise Problems* are included at the end of each chapter. The end-of-chapter problems must also be solved by the student/reader, followed by the verification of their respective answers; the answers are provided at the end of the book. Based on my long-term associations with academia and industry, I anticipate that this volume proves a useful literature for both engineering students and practicing engineers. In particular, engineers/ machine shop managers may greatly benefit from the 200 worked examples/ calculations.

Jeddah, Saudi Arabia   Zainul Huda

Acknowledgments

Thanks to God who blessed me with wisdom that enabled me to write this book. I would like to appreciate my wife for her extended cooperation during the writing of this book. I would like to acknowledge Jamal H. Qadri, PhD (Materials Engineering), President, Solid Solutions International LCC, Sahuarita, Arizona, USA, for reviewing the *classification chart of material failures* for Chap. 1. I am thankful to Frank A. Kulacki, PhD (Mechanical Engineering), Professor, Mech. Eng. Dept., University of Minnesota, USA, for his valuable comments in improving the manuscript of this book. I am also indebted to Khalid Munawwar, PhD (Aeronautical & Space Engineering), Professor, Faculty of Engineering, King Abdulaziz University, Saudi Arabia, for his assistance in developing a high-quality diagram for Chap. 8.

I am grateful to Atiq-ur-Rab Siddique, BS (Mechanical Engineering), MS (Metallurgical Engineering), Senior Lab Engineer, Department of Mechanical Engineering, NED University of Engineering & Technology, Karachi, Pakistan, for reviewing the calculations throughout the book. I gratefully appreciate Siddiqa Kashif d/o Syed Kashif Noman, resident of Karachi, Pakistan, for reviewing the citations of figures and equations in some chapters. The assistance in developing a good quality diagram for Chap. 6 by Abdul Hadi s/o Shoaib Ahmed Khan, resident of Jeddah, Saudi Arabia, is also appreciated. I would like to acknowledge Jack Kane, CEO, EPI, Inc., USA, for his permission to use the photograph of a shaft that failed by fatigue for Chap. 13. I am also grateful to Greg L. Tomei, Manager, Marketing Content, Babcock & Wilcox Company, USA, for his permission to use a photograph (of a boiler's superheater tube that failed by creep) for Chap. 14.

Contents

Part I Materials: Deformation, Testing, and Strengthening

1 Introduction 3
  1.1 What Is Mechanical Behavior of Materials? And Why Study It? . . 3
  1.2 Deformation Behaviors 4
    1.2.1 Deformation and Its Classification. 4
    1.2.2 Time-Independent Deformation – *Elastic/Plastic*
            Deformation 4
    1.2.3 Time Dependent Deformation – *Creep* 6
  1.3 Materials' Failure – *Classification and Disasters* 7
    1.3.1 Fracture and Failure 7
    1.3.2 Classification of Material Failures 7
    1.3.3 Ductile and Brittle Failures 7
    1.3.4 Ductile-Brittle Transition (DBT) Failure 10
    1.3.5 Fatigue Failure 11
  1.4 Materials Selection in Design 11
  1.5 The Factor of Safety in Design 12
  Questions and Problems 13
  References 14

2 Physics of Deformation 15
  2.1 Significance of Crystallography in Deformation Behavior 15
  2.2 Crystallography 15
    2.2.1 Crystalline Solids and Crystal Systems 15
    2.2.2 Crystal-Structure Properties 17
    2.2.3 Crystal Structures of Metals 17
    2.2.4 Miller Indices 20
    2.2.5 Crystallographic Directions 20
    2.2.6 Crystallographic Planes 21

2.3   Crystal Imperctions – *Dislocations*   22
    2.3.1   Cystal Defects   22
    2.3.2   Dislocations   23
2.4   Deformation Mechanisms – *Dislocation Movement*   24
    2.4.1   Deformation by Slip.   24
    2.4.2   Deformation by Twinning   26
2.5   Plastic Deformation – Cold Working/Rolling   27
2.6   Deformation in Single Crystals – *Schmid's Law*   28
2.7   Calculations – *Worked Examples*   30
Questions and Problems   36
References   36

3   Mechanical Testing and Properties of Materials   39
3.1   Material Processing and Mechanical Properties   39
    3.1.1   Relationship Between Processing and Properties   39
    3.1.2   Mechanical Properties/Behaviors   39
3.2   Shear Stress and Shear Modulus   40
3.3   Tensile Testing and Tensile Properties   42
    3.3.1   Tensile Testing   42
    3.3.2   Tensile Mechanical Properties   43
3.4   Elastic Mechanical Properties   46
3.5   Hardness Testing and Hardness of Materials   47
    3.5.1   Hardness and its Testing   47
    3.5.2   Brinell Hardness Test.   47
    3.5.3   Rockwell Hardness Test.   48
    3.5.4   Vickers Hardness Test   50
    3.5.5   Knoop Hardness Test.   51
    3.5.6   Microhardness Test   51
    3.5.7   Hardness Conversion   52
3.6   Impact Toughness – *Impact Energy*   52
    3.6.1   Impact Testing   52
    3.6.2   The Analysis of Impact Testing   53
3.7   Fatigue and Creep Behaviors   53
3.8   Calculations – *Worked Examples*   54
Questions and Problems   59
References   61

4   Strengthening Mechanisms in Metals/Alloys   63
4.1   Strengthening Mechanisms – *Importance and Basis*   63
4.2   Grain-Boundary Strengthening   64
    4.2.1   The Evolution of Grained Microstructure   64
    4.2.2   Grain-Boundary Strengthening – Hall-Petch
                Relationship   65
4.3   Strain Hardening   67
4.4   Solid-Solution Strengthening.   69
4.5   Precipitation Strengthening   70

4.6   Dispersion Strengthening – *Mechanical Alloying*   72
4.7   Calculations – *Worked Examples (Solved Problems)*   72
Questions and Problems   77
References.   79

5   **Materials in Engineering**   81
5.1   Materials and Engineers.   81
5.2   Classification of Materials in Engineering.   81
5.3   Metals and Alloys   82
5.4   Cast Irons.   83
       5.4.1   Characteristics and Applications of Cast Irons   83
       5.4.2   Types of Cast Irons   83
       5.4.3   Mechanical Properties of Cast Irons   85
5.5   Steels   86
       5.5.1   Steels' Definition, Classification, and Designation
                Systems   86
       5.5.2   Carbon Steels.   87
       5.5.3   Alloy Steels   88
5.6   Non-Ferrous Metals and Alloys.   91
       5.6.1   Aluminum and its Alloys.   91
       5.6.2   Copper and its Alloys.   94
       5.6.3   Nickel and its Alloys   95
       5.6.4   Titanium and its Alloys   95
5.7   Ceramics and Glasses   96
       5.7.1   Introduction to Ceramics   96
       5.7.2   Traditional Ceramics   96
       5.7.3   Advanced Ceramics.   97
5.8   Polymers and Plastics   97
       5.8.1   Introduction to Polymers and Plastics   97
       5.8.2   Plastics – *Mechanical Behaviors and Applications*   98
5.9   Composite Materials   100
5.10   Semiconductors and Advanced Materials   101
5.11   Calculations – *Worked Examples/Solved Problems*   101
Questions and Problems   104
References.   106

Part II   Stresses, Strains, and Deformation Behaviors

6   **Stress-Strain Relations and Deformation Models**   109
6.1   True Stress and True Strain   109
6.2   Stress-Strain Relationships – *Young's–, Tangent–*, and *Plastic
       Moduli*   111
6.3   Stress-Strain Relationship in Strain Hardening   112
6.4   Elastic and Plastic Deformation Modles – *Yield Criteria*   113
6.5   Calcualtions – *Worked Examples*.   114
Questions and Problems   118
References.   118

7   Elasticity and Viscoelasticity   119
   7.1   Elastic Behavior of Materials.   119
      7.1.1   Elasticity and Elastic Constants.   119
      7.1.2   Anisotropic and Isotropic Materials.   119
   7.2   Poisson's Ratio.   120
   7.3   Resilience.   121
   7.4   Generalized Hook's Law – *Hook's Law for Three Dimensions*. . . .   122
   7.5   Bulk Modulus – *Relationship Between the Elastic Constants*.   124
      7.5.1   Elastic Constants – *E, G, and B*.   124
      7.5.2   Derivation of Expression for the Bulk Modulus   125
      7.5.3   Relationships Between the Elastic Constants
           and the Poisson's Ratio   128
   7.6   Thermal Effects on Elastic Strains.   128
   7.7   Viscoelasticity   129
   7.8   Calculations – *Worked Examples*.   130
   Questions and Problems   140
   References.   142

8   Complex/Principal Stresses and Strains   143
   8.1   Complex Stresses.   143
      8.1.1   Technological Importance of Complex and Multiple
           Stresses.   143
      8.1.2   What Is a Complex Stresses Situation?   143
   8.2   The State of Plane Stress – *Axes Transformation*   144
      8.2.1   Analyses for Direct and Shear Stresses   145
   8.3   Principal Stresses.   147
   8.4   Mohr's Circle – *Graphical Representation of Stresses*   149
   8.5   Generalized Plane Stress – *The Presence of σ_z in the
      Plane Stress*   149
   8.6   Principal Stresses and the Maximum Shear Stress – *3D
      Consideration*.   150
   8.7   Complex Strains – *Principal Strains in 3 Directions*.   152
   8.8   Calaculations – *Worked Examples*   153
   Questions and Problems   162
   References.   163

9   Plasticity and Superplasticity – *Theory and Applications*   165
   9.1   Plasticity – *Design and Manufacturing Approaches*   165
   9.2   The Stress-Strain Curve and Plasticity   165
   9.3   Plastic Instability in Uniaxial Loading   166
   9.4   Bauschinger Effect.   168
   9.5   Bending of Beams – *Plastic Deformation*   169
      9.5.1   Deriving Expressions for the Curvature and the Radius
           of Curvature.   169
      9.5.2   Symmetrical Bending and the Longitudinal Strain
           in Simply Supported Beams   170

9.6   Application of Plasticity to Sheet Metal Forming   172
        9.6.1   Principal Strain Increments in Uniaxial Loading.   172
        9.6.2   Plane Stress Deformation in Sheet Metal Forming   173
9.7   Hydrostatic Stress and the Deviatoric Stresses   174
9.8   Levy-Mises Flow Rule and Relation Bewteen α and β   175
9.9   Effective Stress and Effective Strain   177
9.10   Superplasticity   177
9.11   Calculations – *Worked Examples*   178
Questions and Problems   186
References.   188

10   Torsion in Shafts   189
10.1   Torsion/Stresses in Shafts   189
        10.1.1   Torsional Shear Stress in a Shaft   189
        10.1.2   Twist and Shear Strain   190
        10.1.3   Power and Torque Relationship and Shaft Design   191
        10.1.4   Torsional Flexibility and Stiffness   191
10.2   Calcualtions – *Worked Examples*   192
Questions and Problems   198
References.   198

Part III   Failure, Design, and Composites Behavior

11   Failures Theories and Design   201
11.1   Failures and Theories of Failure   201
11.2   Maximum Principal Normal Stress Theory or *Rankine Theory* . .   202
11.3   Maximum Shear Stress Theory of Failure *or Tresca Theory*   202
        11.3.1   Theoretical Aspect of Tresca Theory   202
        11.3.2   Design Application of Tresca Theory   204
11.4   Von Mises Theory of Failure   206
        11.4.1   Theoretical Aspect of von-Mises Theory   206
        11.4.2   Design Aspect of von-Mises Theory of Failure   206
11.5   Calcualtions – *Worked Examples*   208
Questions and Problems   212
References.   213

12   Fracture Mechanics and Design   215
12.1   Engineering Failures and Evolution of Fracture Mechanics   215
12.2   Griffith's Crack Theory   216
12.3   Stress Concentration Factor   218
12.4   Loading Modes in Fracture Mechanics   220
12.5   Stress Intensity Factor (K), K_c, and K_{IC}   221
12.6   Design Philosophy   223
        12.6.1   What Is the Design Philosophy of Fracture
                 Mechanics?   223

12.6.2   Application of Design Philosophy to Decide
         whether or Not a Design Is Safe.   223
12.6.3   Application of Design Philosophy to Material
         Selection.   224
12.6.4   Application of Design Philosophy to Design
         of a Testing/NDT Method.   224
12.6.5   Application of Design Philosophy to the Determination
         of Design Stress   224
12.7   Calculations – *Worked Examples*.   224
Questions and Problems   229
References.   231

13   Fatigue Behavior of Materials   233
13.1   Fatigue Failure – *Fundamentals*   233
13.2   Stress Cycles   233
       13.2.1   Types of Stresses and Stress Cycles.   233
       13.2.2   Stress Cycle Parameters   235
13.3   Fatigue Testing – *Determination of Fatigue Strength
       and Fatigue Life*.   236
13.4   Goodman's Law.   238
13.5   Techniques in Designing against Fatigue Failure   238
13.6   Miner's Law of Cumulative Damage.   239
13.7   Fatigue Crack Growth Rate and Computation of Fatigue Life . . .   240
13.8   Calculations – *Worked Examples*.   242
Questions and Problems   249
References.   251

14   Creep Behavior of Materials   253
14.1   Creep Deformation and Failure   253
14.2   Creep Testing and Creep Curve   253
14.3   Factors Controlling Creep Rate   256
14.4   Larson-Miller Parameter (*LMP*)   257
14.5   Creep-Limited Alloy Design   258
14.6   Calculations – *Worked Examples*.   258
Questions and Problems   264
References.   265

15   Mechanical Behavior of Composite Materials   267
15.1   Composite Materials, Classification, and Applications   267
15.2   Mechanical Behavior of Fibrous Composites   269
       15.2.1   General Mechanical Behavior of Fibrous Composites. . .   269
       15.2.2   Behavior of Unidirectional Continuous Fiber
                Composite under Longitudinal Loading.   270
       15.2.3   Stiffness of Unidirectional Continuous Fiber
                Composite under Transverse Loading   272

15.2.4   Poisson's Ratio of Composite Material   273
15.2.5   Shear Modulus of Fibrous Composite Materials   273
15.3   Mechanical Behavior of Particulate Composites.   273
15.4   Calculations – Worked Examples   274
Questions and Problems   280
References.   281

Answers to Problems   283

Index.   287

Author's Biography

Dr. Zainul Huda (ORCID ID: 0000-0002-3433-4995) is a professor of manufacturing technology at the Mechanical Engineering Department, King Abdulaziz University (KAU), Jeddah, Saudi Arabia. His teaching interests include: fracture mechanics, manufacturing technology, metallurgy, and mechanical behavior of materials. He is a principal investigator (*PI*) in an SAR 35,000 research project on boiler super-heater tubes material. He (as a *PI*) has attracted nine research grants, worth total of US$ 0.16 Million. He has been working as a full professor in reputed universities (including the University of Malaya, King Abdulaziz University, and King Saud University) since February 2007. Prof. Huda possesses 40 years' professional experience in materials, manufacturing, and mechanical engineering. He is a professional engineer (PE) registered under Pakistan Engineering Council. He has worked as plant manager, development engineer, metallurgist, and graduate engineer in various manufacturing companies, including Pakistan Steel Mills Corporation Ltd. Prof. Huda earned his PhD in materials technology from Brunel University, London, UK, in 1991. He is also a postgraduate in manufacturing engineering. He obtained BEng (metallurgical engineering) from the University of Karachi, Pakistan in 1976.Prof Huda is the author/co-author of 130 publications, which include 8 books and 58 peer-reviewed international journal/periodical articles (27 ISI-indexed journal papers) in the fields of materials, manufacturing, and mechanical engineering published by reputed publishers from USA, Canada, UK, Germany, France, Switzerland, Pakistan, Saudi Arabia, Malaysia, and Singapore. He has been cited over 850 times in Google Scholar. His author's h-index is 14 (*i*10: 18). He has also successfully completed 20+ industrial consultancy/R&D projects in the areas of failure analysis and manufacturing in Malaysia and Pakistan. He is the developer of Toyota Corolla cars' axle-hub's heat treatment procedure first ever implemented in Pakistan (through Indus Motor Co/Transmission Engineering Industries Ltd, Karachi). Prof. Huda has delivered guest lectures in the United Kingdom, South Africa, Saudi Arabia, Pakistan, and Malaysia. He is a member of prestigious societies, including Institution of Mechanical Engineers (*IMechE*), London, UK; Canadian

Institute of Canadian Institute of Mining, Metallurgy and Petroleum (CIM), Canada; and lifetime member of Pakistan Engineering Council. Prof. Zainul Huda is one of the world's top 20 scholars in the field of materials and manufacturing (see www. scholar.google.com).

Symbol Nomenclature

a	Lattice parameter, m
a	Half crack length of a central crack, m
a	Length of an edge crack, m
a_0	Initial crack size, m
a_f	Final crack size, m
a_c	Critical length of crack, m
A_0	Original cross-sectional area, m²
A	Instantaneous cross-sectional area, m²
A_{CW}	Cross-sectional area after cold work, m²
A_s	Area over which the shear force acts, m²
APF	Atomic packing factor
B	Bulk modulus
c	Solute concentration in the solid solution
C	Material's constant; $LMP = \dfrac{T}{1000}\,(C + log\ t)$
CE	Carbon equivalent (wt %)
d	Draft, m
d_1	Indentation diagonal 1, mm (in Vickers hardness test)
d_2	Indentation diagonal 2, mm (in Vickers hardness test)
d	Mean of d_1 and d_2, mm
d	Average grain diameter, m (in Hall-Petch relationship)
d_p	Particle diameter, m
D_o	Original diameter, m
D	Diameter of solid shaft, m
D	Diameter of indenter, mm (in Brinell hardness test)
D_i	Diameter of indentation, mm (in Brinell hardness test)
D_x	Diameter of extrudate, m (a viscoelasticity effect)
D_d	Diameter of die orifice, m (a viscoelasticity effect)
D_s	Inter-particle spacing in a particulate composite, m
DP	Degree of polymerization
2D	2-dimensional

3D	3-dimensional
E	Young's modulus or modulus of elasticity, Pa
E_{cl}	Young's modulus of the composite under longitudinal loading, Pa
E_f	Young's modulus of fiber, Pa
E_m	Young's modulus of matrix, Pa
E_1	Young's modulus measured in direction 1
E_2	Young's modulus measured in direction 2
f	Torsional flexibility, N•m/rad
F	Tensile force (in tensile testing), N
F	Applied load (in hardness testing), kgf
F	Restoring force (in Hook's law), N
F_c	Force acting on composite, N
F_f	Force acting on fibers, N
F_m	Force on matrix, N
$F0$	Minor load (in Rockwell hardness test), N
$F1$	Major load (in Rockwell hardness test), N
F_{max}	Maximum force, N
Fs	Shear force, N
FoS	Factor of safety
g	Gravitational acceleration, $g = 9.81$ m•s^{-2}
G	Shear modulus or modulus of rigidity, Pa
G_{12}	Shear modulus of composite material, Pa
G_f	Shear modulus of fibers, Pa
G_m	Shear modulus of matrix, Pa
h	Permanent increase in the depth of penetration (in Rockwell hardness test)
h	Initial height of the pendulum, m (in impact testing)
h'	Final height of the pendulum, m (in impact testing)
H	Plastic modulus, Pa
J	Polar 2nd moment of inertia, m^4
K	The constant in Hall-Petch relationship
K	Tangent modulus or strength coefficient, Pa
K	Stress intensity factor, MPa-$\sqrt{m}$
K_{min}	Stress intensity factor at stress level σ_{min}, MPa-$\sqrt{m}$
K_{max}	Stress intensity factor at stress level σ_{max}, MPa-$\sqrt{m}$
ΔK	Stress intensity factor range, $\Delta K = K_{max} - K_{min}$
K_I	Stress intensity factor in *mode I* loading
K_c	Critical stress intensity factor, MPa-$\sqrt{m}$
K_{IC}	Plain strain fracture toughness, MPa-$\sqrt{m}$
K_s	Static stress concentration factor
K_f	Fatigue stress concentration factor
k	The reciprocal of lattice parameter ($k = \dfrac{1}{a}$)
k	Curvature, m^{-1} $\left(k = \dfrac{1}{\rho} \right)$ (in bending of beam)

k	Torsional stiffness
l	Gage length at any instant of tensile testing
l_c	Critical fiber length, m
L	The longest diagonal of indentation, mm (in Knoop hardness test)
L	Length of the shaft, m
L_o	Original length, m
L_f	Final length, m
LMP	Larson-Miller parameter
m	Mass of the pendulum, kg (in impact testing)
m	Strain rate sensitivity index (in superplasticity)
m	Repeat unit molecular weight, amu
M	Number-average molecular weight, amu
M	Bending moment, N.m
n	Number of dislocations
n	Strain hardening exponent or work-hardening exponent
n	Material's constant for the expression: $\dot{\varepsilon} = C_1\,\sigma^n$
n_1	Number of applied cycles at the stress level σ_1
n_2	Number of applied cycles at the stress level σ_2
N	Number of cycles
N_f	Fatigue life or number of cycles to failure
N_1	Fatigue life at stress level σ_1
N_2	Fatigue life at stress level σ_2
N_a	Number of atoms per unit cell
N_A	Nodules count per unit area, nod/m^2 (in ductile cast iron)
N_V	Nodules count per unit volume, nod/m^3 (in ductile cast iron)
N_{rpm}	Rotational speed of the shaft, rev/min
P	Test force, kgf (in Knoop hardness test)
P	Power of the motor, W
q_n	Notch sensitivity factor
Q_c	Activation energy for creep, J•mol^{-1}
r	Distance along the radius of the shaft, m
r_s	Swell ratio (in die swell/viscoelasticity)
R	Universal gas constant, R = 8.3145 J $(mol \bullet K)^{-1}$
R	Atomic radius, m
R	A constant ($R = 100$ for diamond indenter, and $R = 130$ units for steel ball indenter)
R_a	Amplitude ratio
R_s	Stress ratio
S_e	*Endurance limit* or Fatigue limit, Pa
t	Thickness, m
t_0	Original thickness, m
t_f	Final thickness, m
Δt	Time duration, s
t	Stress-rupture time or creep life, h
T_0	Original temperature, ºC

T	Final temperature, °C
T	Temperature, K
T	Torque, N•m (in torsion in shaft)
T_m	Melting temperature, °C or K
U_i	Impact energy of the specimen, J (in impact testing)
U_r	Resilience, J/m^3
U_0	Potential energy of plate without a crack, J
U	Potential energy of the cracked plate, J
V	Volume, m^3
V_a	Volume of an atom, m^3
V_{uc}	Volume of the unit cell, m^3
V_f	Volume fraction of fibers, m^3
V_m	Volume fraction of matrix, m^3
V_p	Volume fraction of particles, m^3
x	Distance moved, m (in Hook's law)
x	Distance along x-axis
x	Transverse displacement, m
y	Deflection, m (in bending of beam)
y	Original length, m
X_c	Weight fraction of crystalline material in a polymer sample
Y	Geometric factor in fracture mechanics equation

Greek Symbols

α	Coefficient of thermal expansion, 1/°C
α	Stress ratio
α	Parameter on the size distribution of nodules in ductile cast iron
β	Strain ratio
γ	Shear strain
γ_{xy}	Shear strain on xy-plane
γ_{yz}	Shear strain on yz-plane
γ_{zx}	Shear strains on zx-plane
γ_1	Principal shear strain in direction 1
γ_2	Principal shear strain in direction 2
γ_3	Principal shear strain in direction 3
γ_s	Specific surface energy, J/m^2
a_γ	Lattice parameter of γ (FCC) matrix, nm
$a_{\gamma'}$	Lattice parameter of gamma-prime (γ') phase, nm
b	Burger's vector
θ	Angle of inclination, rad.
θ	Angle of twist or angle of rotation, rad.
θ_n	Angle of rotation to the coordinate axes for the principal normal stresses
θ_s	Angle of rotation to the coordinate axes for the principal shear stresses

θ'	Intermetallic compound: $CuAl_2$ (in precipitation strengthening)
σ	Applied (tensile) stress, Pa
σ_c	Compressive stress, Pa
σ_c	Critical stress – the stress to cause fracture, Pa
σ_0	Yield strength of single crystal, Pa (in Hall-Petch relationship)
σ_0	Nominal applied stress to a cracked plate, Pa
σ_m	Mean stress, Pa
σ_m	Maximum stress at the crack tip, Pa
σ_{all}	Allowable stress or working (design) stress, Pa
σ_{fail}	Failure strength, Pa
σ_{ys}	Yield strength, Pa
σ_{ut}	Ultimate tensile strength or tensile strength, Pa
σ_{flow}	Flow stress, Pa
σ_{eng}	Engineering stress, Pa
σ_{true}	True stress, Pa
σ_x	Normal stress in x-direction, Pa
σ_y	Normal stress in y-direction, Pa
σ_z	Normal stress in z-direction, Pa
σ_h	Hydrostatic stress, Pa $\left(\sigma_h = \dfrac{\sigma_x + \sigma_y + \sigma_z}{3} \right)$
σ_1	Principal normal stress along axis 1
σ_2	Principal normal stress along axis 2
σ_{max}	Maximum direct stress, Pa ($\sigma_{max} = \sigma_1$)
σ_{min}	Minimum direct stress, Pa ($\sigma_{min} = \sigma_2$)
$\sigma_{\tau p}$	Normal stress accompanying the principal shear stresses, Pa
σ_r	Stress range, Pa
σ_a	Stress amplitude, Pa
σ'	Deviatoric stress, Pa
$\bar{\sigma}$	*Effective stress, Pa*
σ_f	Tensile strength of fiber, Pa
ε	Strain
ε_{eng}	Engineering strain
ε_{true}	True strain
ε_e	Elastic strain
ε_{yield}	Elastic strain taken to yielding
$\varepsilon_{lattice}$	Lattice misfit strain
ε_p	Plastic strain
ε_x	Normal strain in x-direction
ε_y	Normal strain in y-direction
ε_z	Normal strain in z-direction
ε_v	Volumetric strain ($\varepsilon_v = \varepsilon_x + \varepsilon_y + \varepsilon_z$)
ε_1	Principal normal strain in direction 1
ε_2	Principal normal strain in direction 2
ε_3	Principal normal strain in direction 3

$\varepsilon_{\gamma 3}$	Normal strain accompanying the principal shear strains
ε_T	Strain caused by the rise in temperature
$\varepsilon_{x(T)}$	Normal strain in x-direction caused by rise in temperature
$\varepsilon_{y(T)}$	Normal strain in y-direction caused by rise in temperature
$\varepsilon_{z(T)}$	Normal strain in z-direction caused by rise in temperature
$\dot{\varepsilon}$	Strain rate, s^{-1}
$\Delta\varepsilon$	Strain increment
$\bar{\varepsilon}$	*Effective strain*
$\dot{\varepsilon}_{(FG)}$	*Creep* strain rates in fine grained (FG) superalloy
$\dot{\varepsilon}_{(CG)}$	Creep strain rate in coarse grained (CG) superalloy
ε_c	Strain in composite
ε_f	Strain in fiber
ε_m	Strain in matrix
δl	Elastic elongation, m
ΔL	Axial elongation, m
ΔT	Rise in temperature, °C
ds	Arc length, m (in bending of beam)
$d\varepsilon_1$	Principal strain increment in direction 1
$d\varepsilon_2$	Principal strain increment in direction 2
$d\varepsilon_3$	Principal strain increment in direction 3
λ	Angle between the tensile axis and the slip direction, rad.
ψ	Angle between the tensile axis and normal to slip plane, rad.
τ	Shear stress, Pa
τ_c	Adhesive bond strength, Pa
τ_{crss}	Critical resolved shear stress, Pa
τ_r	Resolved shear stress, Pa
τ_{xy}	Shear stress on xy-plane, Pa
τ_{yz}	Shear stress on yz-plane, Pa
τ_{zx}	Shear stress on zx-plane, Pa
τ_{max}	Maximum shear stress, Pa
τ_{min}	Minimum shear stress, Pa
τ_1	Principal shear stress on axis 1, Pa $\left(\tau_1 = \dfrac{|\sigma_2 - \sigma_3|}{2}\right)$
τ_2	Principal shear stress on axis 2, Pa $\left(\tau_2 = \dfrac{|\sigma_1 - \sigma_3|}{2}\right)$
τ_3	Principal shear stress on axis 3, Pa $\left(\tau_3 = \dfrac{|\sigma_1 - \sigma_2|}{2}\right)$
$\tau_{max,\ crit}$	Greatest maximum shear stress, Pa
ΔL	$L_f - L_o$ = Change in length, m
ΔL	Axial elongation, m
ρ_D	Dislocation density, dislocations/m^2
ρ	Total density of polymer sample, kg/m^3
ρ	Radius of curvature, m (in bending of beam)

ρ_t Radius of curvature at crack tip, m

ρ_a Density of the fully amorphous polymer, kg/m^3

ρ_c Density of the fully crystalline polymer, kg/m^3

ν Poisson's ratio

ν_{12} Poisson's ratio of composite material

ν_f Poisson's ratio of fiber

ν_m Poisson's ratio of matrix

δD Lateral contrition, m

δ_c Deformation in composite, m

δ_m Deformation in matrix, m

δ_f Deformation in fibers, m

Part I
Materials: Deformation, Testing, and Strengthening

Chapter 1
Introduction

1.1 What Is Mechanical Behavior of Materials? And Why Study It?

What Is Mechanical Behavior of Materials? *Mechanical behavior of materials* refers to the study of deformation and fracture in materials. The study of this field requires adequate knowledge of materials in engineering, including the influence of microstructure on mechanical properties and behavior. It is also important to study the response of a material to applied load through stress-strain relationships (Meyers and Chowla, 2008; Gere and Goodno, 2012). The main topics in this area include: physics of deformation (*e.g.* crystal structures, dislocation movement, deformation by slip, etc.), mechanical testing and properties of materials, strengthening mechanisms, complex stresses and strains, stress-strain relations, deformation models, elasticity & viscoelasticity, plasticity (especially its application to sheet metal forming), theories of failure, fracture mechanics, fatigue behavior of materials, failure and composite design, and creep deformation of materials.

Why Study Mechanical Behavior of Materials? The mechanical behavior of materials is crucial from both safety and economic point of views. Today's modern society heavily relies on machines and structures. In order to design a machine, a vehicle, or a structure that are safe, reliable, and economical, it is important to efficiently use materials that assure us that failure will not occur. It is therefore mandatory for a mechanical/materials engineering student to gain a sound knowledge of the mechanical behavior of materials, specifically such topics as deformation, failure, fatigue, and fracture mechanics. In metal forming industrial practice, a reasonable plastic deformation in the work-piece is desired to obtain the required geometry in the product." In order to avoid fracture, it is necessary to limit *deformation* (*e.g.* stretching, bending, twisting, etc.) of the component. In particular, crack growth must be strictly limited to avoid *fracture* in brittle materials.

Z. Huda, *Mechanical Behavior of Materials*, Mechanical Engineering Series,
https://doi.org/10.1007/978-3-030-84927-6_1

1.2   Deformation Behaviors

1.2.1   Deformation and Its Classification

When a sufficient load is applied to a metal or other structural material, it will cause the material to change shape. This change in shape is called deformation; which may be either time dependent or time independent deformation (see Fig. 1.1). The two types of deformation are briefly explained in the following sub-sections.

1.2.2   Time-Independent Deformation – Elastic/ Plastic Deformation

A time-independent deformation may be either elastic or plastic deformation.

1.2.2.1   Elastic Deformation

Elastic deformations are reversible; here the elastic strain energy is released upon unloading. It means that on release of the load (stress < yield strength), the material returns to its original shape instantaneously. Elastic deformations are essential for satifactory functioning of some machine elements, such as in a mechanical spring (see Fig. 1.2); however this beneficial effect is valid only up to a limit – *elastic limit* (see Chaps. 3 and 8).

It is evident in Fig. 1.2 that the spring is stretched on application of force or load, and it returns to its original position on release of the load. This mechanical behavior is referred to as *elasticity*. Additionally, the extension in the spring is proportional to the applied force *i.e.* the deformation in spring obeys *Hook's law* – the force needed to extend or compress a spring by some distance (x) is proportional to the distance, up to the elastic limit. Mathematically,

$$F = -kx \qquad (1.1)$$

where F is the restoring force *i.e.* force the spring exerts on the object that originally applied the force to the spring, N; x is the extension or compression, m; and k is the

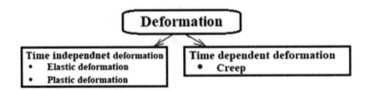

Fig. 1.1 Classification of deformation

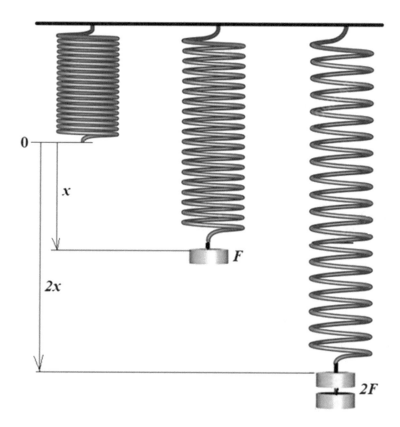

Fig. 1.2 Elastic deformation in a mechanical spring – Hook's law

spring constant or stiffness of the spring, N/m (see Example 1.1). The negative sign in Eq. 1.1 indicates that the restoring force F is in the opposite direction of the force that is stretching or compressing the spring. Elastic deformation/elasticity is discussed in detail in Chap. 7.

Example 1.1 Calculating the Spring Constant (k) by using Hook's law A car will have a mass of 800 kg, and there will be four shock absorbers, each 45-cm long, to work with. Estimate the strength of the shock-absorber by calculating its spring constant.

Solution Assuming these shock absorbers use springs, each one has to support a mass of at least 200 kg $\left(\dfrac{800\, kg}{4} = 200\, kg \right)$. Taking extension in spring = length of the spring = 45 cm = 0.45 m = x.

   The gravitational force or weight supported by each shock absorber can be computed by:

$$F = W = mg = 200\text{kg} \times 9.8\,\text{m}/\text{s}^2 = 1960\,\text{N}$$

By using modified form of Eq. 1.1 or Hooke's law we obtain:

$$k = \frac{F}{x} = \frac{1960}{0.45} = 4,355\,\text{N}\,/\,\text{m}$$

Each of the springs used in the shock absorbers must have spring constant of at least 4355 N/m.

1.2.2.2 Plastic Deformation

Plastic deformations are irreversible *i.e.* plastic deformation is permanent. For designers, plasticity refers to material failure; here *plastic deformation* is a permanent change in the physical dimensions or shape of a component that is sufficient for its function to be lost or impaired. Excessive plastic deformations generally result in failure. For example, collapse of a steel bridge during an earthquake is generally attributed to large plastic deformation. Sometimes, even a small plastic deformation can cause malfunctioning of a machine. For example, a mechanical wrist watch, when dropped to a hard surface, may malfunction due to plastic deformation of its gears or other similar component.

In manufacturing (metal forming) context, plasticity is beneficial. For a manufacturing engineer, plasticity refers to formability – ability to be plastically deformed when stressed. Here, plasticity enables a manufacturing engineer to produce parts by metal forming (*e.g.* rolling, forging, extrusion, sheet metal forming, etc.). The physics of plastic deformation is explained in Chap. 2. Plasticity is discussed in detail in Chap. 9.

1.2.3 Time Dependent Deformation – Creep

Creep refers to the progressive deformation of a material at a constant stress, usually at a high temperature (that exceeds 0.4 to 0.5 of the melting temperature, expressed in Kelvin); however creep may also occur at ambient temperature when the stress is applied for a long period of time. Some notable examples of systems that have components experiencing creep include: gas turbine components, nuclear reactors, boilers, and ovens. In nuclear reactors, the metal tubes carrying the fuel may undergo creep failure in response to the pressures and forces exerted on them at high temperatures. Low-temperature creep can occurs in structures at ambient temperatures when the stress is applied for a long duration of time. For example, pre-stressed concrete beams, which are held in compression by steel rods that extend through them, may creep. The creep and stress relaxation in the steel rods eventually leads to a reduction of the compression force acting in the beam; which can result in failure. Creep deformation is discussed in detail in Chap. 14.

1.3 Materials' Failure – *Classification and Disasters*

1.3.1 Fracture and Failure

The fracture of a component or a structure refers to the development of a crack leading to its fragmentation in response to an imposed stress that is static and applied at moderate temperatures. In a broad context, a component or a structure is said to have failed if one or more of the following situations are encountered: (a) the component becomes completely inoperative, (b) it ceases to perform the function for which it was manufactured due to excessive plastic deformation, (c) it fractures, or (d) it becomes unsafe or unreliable due to serious corrosion/wear/material deterioration. In mechanical systems with moving parts, excessive friction and/or contact between two components may result in failure by *wear*. Since engineering failures impose a threat to society, it is important to prevent them through a systematic failure analysis (Booker, et al., 2020). Some historical failures/disasters are discussed in sub-sections 1.3.3 and 1.3.4.

1.3.2 Classification of Material Failures

A polycrystalline material can fail during monotonic loading in one of the following ways: cleavage/inter-granular brittle fracture (fracture along grain boundaries), trans-granular ductile fracture (fracture through grains), fatigue, creep, hydrogen embrittlement, and stress corrosion cracking. The cracking along specific crystallographic planes, is called *cleavage*. Transgranular ductile failure reveals microvoid coalescence (MVC) under a high resolution microscope. These material failure mechanisms are presented in the classification chart, as shown in Fig. 1.3.

It is evident in Fig. 1.3 that all material failures can be classified into three groups: (a) failures based on material characteristics, (b) failures due to design/manufacturing faults, and (c) failures based on material, loading, or environmental conditions. Environmentally assisted cracking (EAC) includes stress corrosion cracking (SCC), hydrogen embrittlement (HE), and liquid metal embrittlement (LME). Most of the various types of failure, shown in Fig. 1.3, are explained in the subsequent sections.

1.3.3 Ductile and Brittle Failures

Based on the material characteristic, a failure may be either ductile or brittle. Ductile failures are associated with gross plastic deformations prior to failure; this is why they do not offer much threat to society. On the other hand, a brittle fracture of a component often causes a sudden serious damage *i.e.* it is catastrophic. A brittle

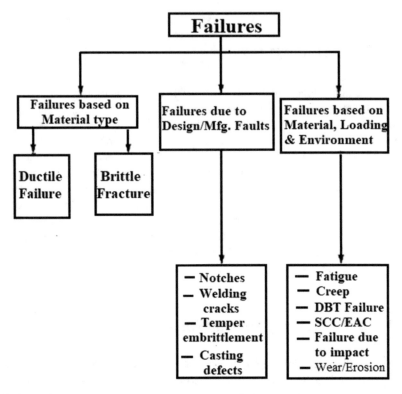

Fig. 1.3 Classification of materials' failures (DBT ductile-brittle transition, SCC stress corrosion cracking, EAC environmental assisted cracking)

fracture usually occurs without warning at a stress far below the yield strength of the material; the latter is explained in Ch. 3. The difference between a ductile and brittle fractures is illustrated in Fig. 1.4.

It is evident in Fig. 1.4 that a ductile fracture usually results in either highly ductile fracture (Fig. 1.4a) or ductile fracture (Fig. 1.4b). A highly ductile fracture involves 100% reduction in area (*e.g.* in lead, thermoplastics, etc), whereas a ductile fracture surface shows a cup-and-cone ductile feature (*e.g.* in mild steel, brass, etc). Both of these types of ductile fracture involve gross plastic deformation prior to failure (see Table 1.1). Figure 1.4(a) shows a slant fracture resulting from shear deformation along a plane inclined 45° to the axis of loading.

Brittle fractures involve no or very little plastic deformation; the fracture surface is smooth and shining (see Fig. 1.4c; see Table 1.1). Materials that exhibit brittle fracture under uniform loading at ambient temperatures and moderate loading rate include: glass, white cast iron, and the like. Brittle fractures generally occur in brittle materials; however they can also occur in ductile materials under certain loading conditions or at very low temperature (see Sect. 1.3.4). For example, an accidental/ impact loading or tri-axial stresses can cause a ductile material to behave in a brittle manner.

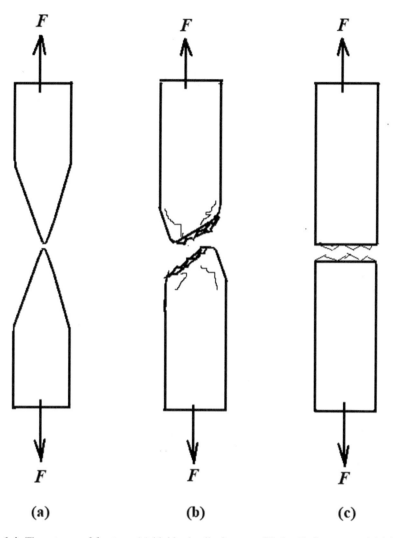

Fig. 1.4 Three types of fracture; (**a**) highly ductile fracture; (**b**) ductile fracture, and (**c**) brittle fracture

Table 1.1 Ductile and Brittle Fractures

#	Ductile fracture	Brittle fracture
1.	It involves a gross plastic deformation of the material prior to failure.	There is no or very little plastic deformation prior to failure.
2.	The material has high impact energy.	The material has a low impact energy.
3.	The fracture surface appears dull and fibrous showing a cup-and-cone.	The fracture surface is usually flat and shining.
4.	Scanning electron microscopy (*SEM*) fractographs usually show spherical dimples due to micro-void coalescence (MVC) in metals.	SEM fractographs of metals show transgranular cleavage and intergranular fracture as important fracture mechanisms.

Fig. 1.5 Drop in impact
energy at low temperatures
(below DBTT) (DBTT
ductile-brittle transition
temperature)

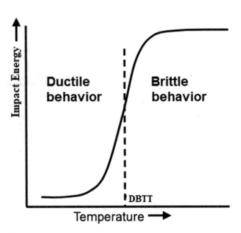

In addition to ductile fracture and brittle fracture, there are often material failures
in mixed mode.

Brittle fractures impose a high impact threat to society. Many reported engineering failures/disasters have been associated with brittle fractures (see Chap. 12,
Sect. 12.1).

1.3.4   Ductile-Brittle Transition (DBT) Failure

Materials with high impact energies exhibit *ductile failure* whereas brittle materials
are characterized by their low impact energy values. Impact energy is the ability of
a material to absorb energy under impact loading (see Chap. 3). The impact energy
of a metal is strongly dependent on temperature. In particular, there is a sharp reduction in impact energy when the temperature falls below the ductile-brittle transition
(DBT) temperature; which is generally below 0 °C. It has been experimentally
shown that metals which exhibit normal ductile fracture at ambient (room) temperature, can fail at low temperatures by a sudden cleavage (brittle) fracture at compara-
tively low stresses (see Fig. 1.5). This phenomenon is called *ductile-brittle transition
(DBT) failure*; and the temperature at which DBT failure occurs is called the ductile-
brittle transition temperature (DBTT) (see Fig. 1.5).

DBT failures are more common at low temperatures. There have been many
instances of failures of metals by unexpected brittleness at low temperatures; for
example the fracture of *Titanic* ship is attributed to DBT failure (see Chap. 12, Sect.
12.1). In general, lower the DBTT of a material, safer is the design against the DBT
failure. Steels (with BCC ferritic structure) fail with brittle fracture along the (100)
plane. For designing such steels against DBT failure, it is important to significantly
reduce the DBT temperature of these steels by minimizing the carbon, phosphorous,
oxygen, nitrogen, and hydrogen contents in the steels.

Fig. 1.6 The three stages
in fatigue failure (Stage I
= crack initiation, (stage
II = crack propagation,
stage III = fatigue failure)

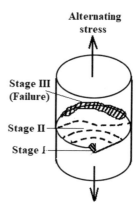

1.3.5   Fatigue Failure

Many engineering situations involve loading of a structure or component with vary-
ing stresses. It is important to know how materials behave under cyclic (dynamic)
loading. When a material is subjected to alternating or cyclic stresses over a long
period of time, it may fail after a number of cycles even though the maximum stress
in any cycle is considerably less than the breaking strength of the material. This
failure is called *fatigue*. Typical examples of components, experiencing fatigue failures, include: rotating shafts, springs, turbine blades, airplane wings, gears, bones,
and the like. There are three main stages in a fatigue failure: (I) ***crack initiation***, the
crack originates at a point of stress concentration (such as a notch) or a metallurgical
flaw (e.g. inclusion); (II) ***crack propagation***, the crack propagates across the part
under cyclic or repeating stresses, and (III) ***crack termination,*** here final fracture
occurs (see Fig. 1.6, Fig.13.1). Fatigue failure is discussed in detail in Chap. 13.

1.4   Materials Selection in Design

Design of a machine or its component involves the specification of its geometric
shape, materials selection, manufacturing method, and other details needed to completely describe the machine, vehicle, structure, or its component. Materials selection in design is a crucial activity since several important factors must be considered
in selecting the right material for a product or its component. These factors mainly
include: (1) material properties, (2) material cost and availability, (3) processing,
and (4) environment. *Material properties* determine the expected level of performance from the material (see Chaps. 3, 4, and 5). *Material cost and availability*
encompasses two consideration: (a) material must be appropriately priced (not nec-
essarily cheap but right), and (b) material must be available (preferably from multiple sources). *Processing* refers to considering how to make the part, for example:

by casting, metal forming, machining, or any other manufacturing process (Huda, 2018). *Environment* refers to the effect that the service environment has on the part, the effect the part has on the environment, and the effect that processing has on the environment. The selection of material in design of a component with a crack is explained in Chap. 12. The scope of this book does not permit this author to discuss in detail *materials selection in design*; the reader is advised to refer to the literature (Ashby, 2010; Huda, 2013).

1.5   The Factor of Safety in Design

A machine component will operate safely as long as the working stress (or the allowable design stress) on the component is less than the failure strength. This safety-in-design concept is expressed by the term *factor of safety* (*FoS*), as follows:

$$FoS = \frac{Failure\ strength}{working\ or\ allowable\ stress} = \frac{\sigma_{fail}}{\sigma_{all}} \tag{1.2}$$

where σ_{fail} is the failure strength of the component, MPa; and σ_{all} is the working stress or the allowable (design) stress on the component during service, MPa (see Example 1.2). The *FoS* is a margin of safety in design. It is obvious from Eq. 1.2 that a design is adequate or safe when *FoS* > 1. On the other hand, when FoS ≤ 1, the design is inadequate. It must be noted here that Eq. 1.2 holds good only in case of uniaxial loading of a component. In case of bi-axial loading, an appropriate failure theory must be applied (see Chap. 11). The *FoS* usually ranges from 1.3 (when material properties and operating conditions are known) to 4 (for untried materials) (Huda and Jie, 2016). For elevator applications, FoS may be as high as 7.

Example 1.2 Designing a Component based its Breaking Strength and the Design Stress The failure strength of a component's material is 1100 MPa. The material properties and operating conditions are known in detail. What should be the design (working) stress for the component? Design the cylindrical component for a tensile load of 2 kN.

Solution For known material properties, $FoS = 1.3$; $\sigma_{fail} = 1, 100$ MPa, F = 2000 N, $\sigma_{all} = ?$, d = ?

By re-writing Eq. 1.2, we obtain:

$$\text{Design stress} = \sigma_{all} = \frac{\sigma_{fail}}{FoS} = \frac{1100}{1.3} = 846\ \text{MPa}$$

$$\text{Stress} = \sigma_{all} = \frac{F}{A}$$

$$A = \frac{\pi}{4} d^2 = \frac{F}{\sigma_{all}}$$

$$\text{Diameter} = d = \sqrt{\frac{4F}{\pi \sigma_{all}}} = \sqrt{\frac{4 \times 2000}{\pi \times 846}} = \sqrt{3} = 1.73 \, \text{mm}$$

Questions and Problems

1.1 Define the term mechanical behavior of materials. Why is the study of mechanical behavior of materials important in engineering?

1.2 (a) Classify various types of deformation and define each type of deformation.
(b) Is plastic deformation always undesirable? Justify your answer.

1.3 (a) Distinguish between the terms *fracture* and *failure*.
(b) Classify the various types of material fracture.
(c) Differentiate between ductile fracture and brittle fracture with the aid of sketches.
(d) What is meant by ductile-brittle transition (DBT) failure of a material? Explain with the aid of a sketch. How can DBT failure be avoided in steels?

1.4 Explain fatigue failure indicating the three stages involved in the failure.

1.5 (a) Define the terms *design* and *materials selection*.
(b) What factor must be considered in materials selection in design? Briefly explain them.

1.6 A truck will have a mass of 7000 kg, and there will be four shock absorbers, each 20-in. long, to work with. Calculate the spring constant of the shock absorber.

1.7 A machine component has a diameter of 0.8 mm. The breaking strength of the material is 1800 MPa. Is it safe to use the component under a tensile load of 2 kN?

1.8 (MCQs). Encircle the most appropriate answers for the following statements/questions.

(a) Which failure results due to time-dependent deformation of a material?
(i) fatigue, (ii) creep deformation , (iii) elastic deformation, (iv) plastic deformation
(b) Which failure results due to cyclic loading of a material?
(i) fatigue failure, (ii) creep failure, (iii) ductile failure, (iv) brittle failure
(c) Which failure involves gross plastic deformation of a material?
(i) DBT failure, (ii) creep failure, (iii) brittle failure, (iv) ductile failure
(d) Which failure occurs at high temperatures?
(i) DBT failure, (ii) creep, (iii) brittle failure, (iv) ductile failure
(e) Which failure is associated with a low impact energy of a material?
(i) fatigue failure, (ii) creep failure, (iii) brittle failure, (iv) ductile failure

  (f) Which failure occurs at very low temperatures in BCC metals?

      (i) DBT failure, (ii) creep, (iii) brittle failure, (iv) ductile failure

  (g) Which failure generally occurs in rotating shafts?

      (i) DBT failure, (ii) creep, (iii) brittle failure, (iv) fatigue failure

  (h) Which failure is catastrophic?

      (i) ductile failure, (ii) creep, (iii) brittle failure, (iv) fatigue failure

  (i) What was the cause of *Titanic* ship failure?

      (i) ductile failure, (ii) creep, (iii) DBT failure, (iv) fatigue failure

  (j) Which failure is indicated by inter-granular fracture as observed under a microscope?

  (k) ductile failure, (ii) creep, (iii) fatigue failure, (iv) brittle failure

References

Ashby MF (2010) Materials selection in mechanical design, 4th edn. Butterworth- Heinemann/ Elsevier, Oxford

Booker NK, Clegg RE, Knights P, Gates JD (2020) The need for an international recognised standard for engineering failure analysis. Eng Fail Anal 110:104357

Gere JM, Goodno BJ (2012) Mechanics of materials, 8th edn. Cengage Learning, Boston

Huda Z (2013) Materials selection in design of structure and engine of supersonic aircrafts. Mater Des 46:552–560

Huda Z (2018) Manufacturing: mathematical models, problems, and solutions. CRC Press, Boca Raton

Huda Z, Jie EHC (2016) A user-friendly approach to the calculation of factor of safety for pressure vessel design. J King Abdulaziz University (JKAU) 27(1):75–81

Meyers MA, Chowla KK (2008) Mechanical behavior of materials, 2nd edn. Cambridge University Press, Cambridge, UK

Chapter 2
Physics of Deformation

2.1 Significance of Crystallography in Deformation Behavior

The study of mechanical behavior of materials involves a great deal of deformation of solids. Plastic deformation occurs as a result of dislocation motion across specified crystallographic plane(s) when shear stresses are applied along the plane in the specified direction. A dislocation is a crystal defect; it is referred to as a line defect. Hence, a sound understanding of crystallography and crystal defects (*e.g.* dislocations) is of paramount importance in explaining the deformation of solids. This chapter, therefore, first presents an overview of crystal structures and defects with particular reference to dislocations. Then the role of crystallographic system (crystallographic plane and direction) and dislocation movements in plastic deformation, is discussed with particular references to slip, twinning, and deformation by slip of single crystals.

2.2 Crystallography

2.2.1 Crystalline Solids and Crystal Systems

What Is a Crystalline Solid? All solid materials may be classified into two groups: (1) amorphous solids, and (2) crystalline solids. An *amorphous solid* has no definite geometric form; it is any non-crystalline solid that does not organize the atoms and molecules in a definite lattice pattern. On the other hand, in a crystalline solid, atoms are arranged in a definite 3-dimensional (3D) geometric pattern with long-term periodicity. Thus, a crystal is a solid in which the atoms take up a highly ordered, definite, geometric arrangement that is repeated 3-dimensionally within the crystal. The crystal structure of quartz (SiO_2) is shown in Fig. 2.1; which is

© The Author(s), under exclusive license to Springer Nature Switzerland AG 2022
Z. Huda, *Mechanical Behavior of Materials*, Mechanical Engineering Series,
https://doi.org/10.1007/978-3-030-84927-6_2

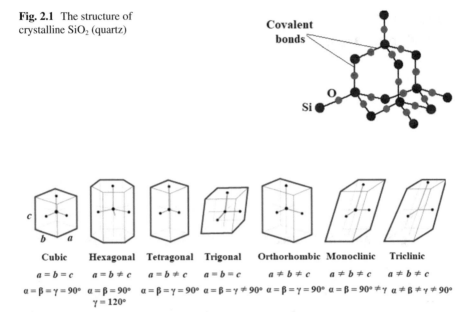

Fig. 2.1 The structure of crystalline SiO_2 (quartz)

Fig. 2.2 The seven crystal systems' structures

similar to a diamond crystal. Crystalline solids are the most abundant (90%) of all naturally occurring and artificially prepared solids. Examples of crystalline solids include: metals, minerals, limestone, diamond, graphite, salts (sodium chloride, potassium chloride, etc), and the like.

Crystal Systems All crystalline solids may be broadly classified into the following seven crystal systems: (a) cubic, (b) hexagonal, (c) tetragonal, (d) trigonal, (e) orthorhombic, (f) monoclinic, and (g) triclinic; the unit cell geometry for each of the seven systems is illustrated in Fig. 2.2. The smallest repeating array of atoms in a crystal is called a *unit cell*. The simplest/most symmetric crystal system is the cubic system; here all edges of the unit cell are equal to each other ($a = b = c$), and all the angles are equal to 90° ($\alpha = \beta = \gamma = 90°$). The tetragonal and orthorhombic systems refer to rectangular unit cells, but the edges are not all equal. In the remaining systems, some or all of the angles are not equal to 90° (Helliwell 2015). The least symmetrical structure is the triclinic system; here no edges are equal and no angles are equal to each other or to 90° (Vainshtein et al. 1995). In the hexagonal system, two edges of the unit cell are equal and subtend an angle of 120°. Hexagonal crystals are quite common both in metals and ceramics; examples of materials with hexagonal crystal system include: zinc, titanium, graphite, and the like.

2.2.2   *Crystal-Structure Properties*

A crystal structure is generally described by the following three properties: (a) number of atoms per unit cell, (b) coordination number (CN), and (c) atomic packing factor (*APF*). The *CN* refers to the number of atoms in direct contact with any given atom in the crystal lattice. The fraction of volume occupied by atoms in a unit cell, is called the atomic packing factor (*APF*).

$$APF = \frac{N_a V_a}{V_{uc}}$$

(2.1)

where N_a is the number of atoms per unit cell, V_a is the volume of an atom, and V_{uc} is the volume of the unit cell (see Example 2.1).

2.2.3   *Crystal Structures of Metals*

Almost all metals are crystalline solids at room temperature; the only exception is mercury. Metals have crystal structures that can be described as body center cubic (*BCC*), face centered cubic (*FCC*), or hexagonal close packed (*HCP*); the three crystal structures are discussed as follows.

2.2.3.1   Body-Centered Cubic (BCC) Structure

The body-centered cubic (*BCC*) unit cell is a cubic structure having atoms at each of the eight corners of the cube plus one atom in the cube-center (see Fig. 2.3a). Since each of the corner atoms is the corner of another cube, the corner atom in each unit cell is shared among eight unit cells. The contribution of each atom to the BCC unit cell is illustrated in Fig. 2.3(b). Examples of metals crystallizing in BCC structure include: alpha-iron (α-Fe), chromium (Cr), niobium (Nb), vanadium (V), molybdenum (Mo), and the like.

Figure 2.3(b) enables us to calculate the number of atoms per unit cell for BCC lattice.

$$\text{No. of atoms per unit cell for } BCC = \left(8 \times \frac{1}{8}\right) + 1 = 2$$

(2.2)

The coordination number for BCC crystal structure can be found as follows. Since the central atom in the BCC unit cell is in direct contact with eight corner atoms, the coordination number for BCC unit cell is: CN = 8 (see Fig. 2.3). It can be shown that the atomic packing factor (APF) for BCC crystal lattice is 0.68 (see Example 2.1).

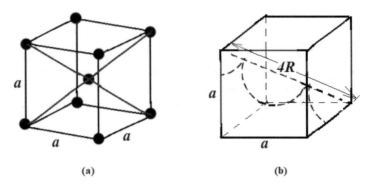

Fig. 2.3 BCC crystal structure; (**a**) BCC unit cell, (**b**) BCC unit cell showing contribution of each atom to the unit cell (R = atomic radius)

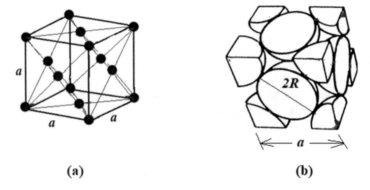

Fig. 2.4 FCC crystal structure's unit cell (a = lattice parameter)

2.2.3.2   Face Centered Cubic (FCC) Structure

The *face-centered cubic* (FCC) unit cell has atoms at each corner of the cube and six atoms at each face of the cube (see Fig. 2.4a). In the FCC structure, each corner atoms contributes one-eighth to the unit cell whereas the atom at each face is shared with the adjacent cell (Fig. 2.4b). Examples of metals crystallizing in FCC structure include: gamma iron (γ-Fe), copper (Cu), nickel (Ni), aluminum (Al), platinum (Pt), and the like.

The face diagonal in an FCC unit cell is equal to $4R$ (see Fig. 2.4b). Thus we can develop a relationship between the atomic radius (R) and the lattice parameter (a) for FCC lattice, as follows:

$$\left(4R\right)^2 = a^2 + a^2 \; \left(by \text{ using Pythagoras theorem}\right)$$
$$4R = \sqrt{2}\, a \tag{2.3}$$
$$a = 2\sqrt{2}R = 2.828\,R$$

It can been shown that there are 4 atoms per unit cell in *FCC* crystal lattice (see Example 2.2). The coordination number of the *FCC* lattice can be deduced as follows. The face centered atom is surrounded by four corner atoms of its own plane atoms; and the adjacent 8 face centered atoms (see Fig. 2.4a). This makes the total number of the nearest neighbors equal to 12. Hence, the coordination number (*CN*) of *FCC* lattice = 12. It can also be shown that the atomic packing factor (*APF*) for *FCC* lattice is 0.74 (see Example 2.1).

The *CN* and *APF* values of the *BCC* with the *FCC* lattices are listed in Table 2.1. For *BCC*, the *CN* and *APF* values are lower than those for FCC. This crystallographic behavior indicates that the *BCC* lattice does not allow the atoms to pack together as closely as the *FCC* lattice does.

2.2.3.3 Hexagonal Close Packed (HCP) Crystal Structure

The HCP unit cell can be visualized as a top and a bottom planes; each plane has 7 atoms thereby forming a regular hexagon around a central atom. In between these two planes, there is a half hexagon of 3 atoms (see Fig. 2.5). In the HCP unit cell, one lattice parameter (*c*) is longer than the other (*a*) such that $c = 1.63\,a$. It is evident

Table 2.1 Crystal structure properties of some metals

Crystal structure	No. of atoms/unit cell (n)	CN	APF	Examples
BCC	2	8	0.68	*Fe*, Cr, Nb, V, Mo*
FCC	4	12	0.74	*Fe†, Al, Ni, Cu, Pt*
HCP	6	12	0.74	Zinc, titanium

Fe* = α-iron at temperatures: 25–910 °C
Fe† = γ-iron at temperatures: 910–1410 °C

Fig. 2.5 The HCP crystal structure; (**a**) HCP unit cell showing contribution of **atoms to the unit cell, (b) HCP crystal lattice**

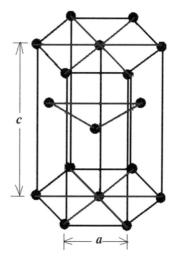

in Fig. 2.5 that each corner atom, in the HCP unit cell, contributes one-sixth to the cell whereas the atom at each face is shared with the adjacent cell; the three middle-layer atoms contribute full to the cell. It can be proved that there are six (6) atoms per unit cell in the HCP lattice (see Example 2.2). Examples of metals crystallizing in HCP structure include: zinc, titanium, and the like (see Table 2.1).

In an HCP crystal lattice, there are 6 nearest neighbors in the same close packed layer (3 in the layer above and 3 in the layer below). This makes the total number of the nearest neighbors equal to 12. Hence, coordination number (*CN*) of HCP lattice is 12. It can be mathematically shown that the atomic packing factor (APF) for HCP lattice is 0.74 (see Examples 2.3 and 2.4). The crystal properties data for BCC, FCC, and HCP are summarized in Table 2.1.

The data in Table 2.1 indicates that both FCC and HCP crystal structures have higher CN and APF values as compared to BCC lattice. It means that FCC and HCP lattices are closed packed structures whereas BCC is not. A close-packed crystal structure ensures the strongest possible metallic bond among atoms in the metal (Douglas and Ho 2006).

2.2.4 Miller Indices

Miller indices are used as the notation system for defining planes and directions in crystal lattices. The specifications of directions and planes within crystalline solids are of great technological importance in explaining the mechanical behavior of crystalline materials. In order to define a crystallographic plane, three integers *l*, *m*, and *n*, known as *Miller indices*, are used. A crystallographic plane is represented by integers (*hkl*), whereas a crystallographic direction is represented by [*hkl*]. By convention, a negative integer is represented by a bar (*e.g.* $-k = \bar{k}$).

2.2.5 Crystallographic Directions

Any vector direction specified by two points in a crystal lattice, is called a *crystallographic direction*. The procedure to define a crystallographic direction, by finding its Miller Indices, is described as follows.

  (I) Locate 2 points lying on the given crystal direction.
 (II) Determine the coordinates of the two points using a right hand coordinate system as defined by the crystallographic axes x, *y*, and *z*.
(III) Obtain the number of lattice parameters moved in the direction of a coordinate axis by subtracting the coordinates of the *tail point* from that of the *head point*.
(IV) Clear fraction and/or reduce the results so obtained to the smallest integers.
 (V) Represent the Miller Indices as [*hkl*].

Fig. 2.6 The Miller
Indices for coordinate axes
in a cubic crystal

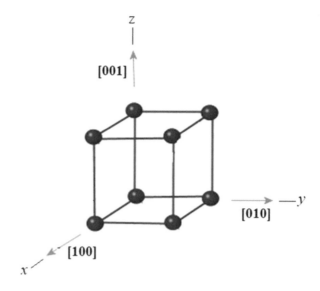

The above-listed set of procedures (I-V), used to find *Miller Indices* for a crystal-lographic direction, are illustrated in Example 2.5. For cubic crystals, the *Miller Indices* are the vector components of the direction resolved along each of the 3 coordinate axes and reduced to the smallest integers *i.e.* [100], [010], and [001] (see Fig. 2.6). The family of the crystallographic directions for the three orthogonal axes, in cubic crystals, is denoted by <100>.

Crystallographic directions are important in explaining the mechanical behavior of materials. For example, in the processing of grain-oriented electrical steels (*GOES*), excellent magnetic properties are achieved in the [100] rolling direction. The *GOES* steel, containing up to 4% silicon, finds application in transformers for power generation (see Chap. 5, Sect. 5.5.3.2).

2.2.6 Crystallographic Planes

2.2.6.1 Procedure to Find Miller Indices for a Crystallographic Plane

The orientation of an atomic plane in a crystal lattice, is referred to as a *crystallo-graphic plane*. In all crystal systems, except the hexagonal system, crystallographic planes are specified by three Miller Indices as (*kkl*). The *indices (hkl)* are the recip-rocals of the fractional intercepts (with fractions cleared) that the plane makes with the three orthogonal crystallographic axes *x*, *y*, and *z*.

In order to define the Miller Indices, the following procedure should be followed: (I) find the fractional intercepts that the plane makes with the crystallographic axes; it means that find how far along the unit cell lengths does the plane intersect the

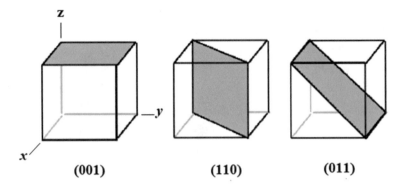

Fig. 2.7 Miller Indices for some crystallographic planes

axis, (II) take the reciprocal of the fractional intercept of each unit length for each axis, (III) clear the fractions, and (IV) enclose these integer numbers in parenthesis (), and designate that specific crystallographic plane within the lattice.

The use of the above-listed steps (I-IV), in determining the *Miller Indices*, is illustrated in Examples 2.6 and 2.7. Some crystallographic planes with Miller Indices are shown in Fig. 2.7.

2.2.6.2   Applications of Crystallographic Planes to Deformation of Solids

Crystallographic planes play an important role in the mechanical behavior of materials. For example, plastic deformation of metals occurs along specified crystallographic planes in specified directions. In FCC metals, deformation by slip can occur along the planes (111), (1 $\bar{1}$ 1), (11 $\bar{1}$), and ($\bar{1}$ 11) (see Sect. 2.4.1). In semiconductor fabrication, it is easier to cleave the silicon wafer along (111) plane than along other crystallographic planes.

2.3   Crystal Imperctions – *Dislocations*

2.3.1   Cystal Defects

Theoretically, (perfect) crystals are defined as 3-dimensional perfectly ordered arrangements of atoms or ions. However, real crystals contain defects or imperfections. This is why the yield strength of real crystals is much lower than the theoretically calculated yield strength of perfect crystals (Huda and Bulpett 2010). *Crystal imperfections* refer to the missing of atoms/ions or misalignment of unit cell in an otherwise perfect crystal. Crystal imperfections or defects can be classified into the following four groups: (1) point defects, (2) line defects, (3) surface defects, and (4) volume defects. A *point defect* occurs at a single lattice point (*e.g.* a vacancy). A *line defect* occurs along a row of atoms (*e.g.* a *dislocation*); or as a planar defect occurring over a 2-dimensional surface in the crystal. A volume defect is generally a *void* in the crystal. In general, crystal imperfections occur either as *point defects* or *line defects*. The *dislocation* line defect plays a pivot role in the plastic deformation of materials. Since the scope of this book does not permit us to discuss all the crystal imperfections, this chapter focuses on the explanation of the formation and movement of dislocations (see Sections 2.3.2.2 and 2.4).

2.3.2   Dislocations

2.3.2.1   What Is a Dislocation?

A *dislocation* is a line defect (crystal imperfection) that occurs along a row of atoms. *Dislocations* cause lattice distortions centered round a line in a crystalline solid; they can be observed by use of transmission electron microscopy (*TEM*) techniques (see Fig. 4.6, Chap. 4). In industrial practice, dislocations are introduced during solidification, plastic deformation, or as a consequence of thermal stresses that result from rapid cooling; this is why almost all crystalline materials contain dislocations (Hull and Bacon 2011). Dislocations weaken the crystal structure along a one-dimensional space. An important parameter in dislocation is *dislocation density*; which is defined as the number of dislocations in a unit volume of a crystalline material (Callister 2007).

2.3.2.2   Types of Dislocation – *Edge-, Screw-, and Mixed Dislocation*

The application of shear stresses to a crystalline solid (*e.g.* in metal forming) results in atomic displacements due to dislocation motion. The direction and amount of the atomic displacement caused by the dislocation, is represented by a vector – called the *Burger's vector*, $\bar{b}$ (see Fig. 2.8). There are three principal types of dislocation in a crystalline solid: (a) edge dislocation, (b) screw dislocations, and (c) mixed dislocation.

An *edge dislocation* results due to the insertion of an extra half plane of atoms in a crystal lattice. An *edge* dislocation has its Burgers vector ($\bar{b}$) directed perpendicular to the dislocation line (see Fig. 2.8a). Under the action of a shear stress, an *edge dislocation* moves in the direction of the Burger's vector. A *screw dislocation* has its Burger's vector ($\bar{b}$) oriented parallel to the dislocation line (see Fig. 2.8b). Under the influence of a shear stress, a *screw dislocation* moves in a direction perpendicular to the Burgers vector. The *mixed dislocation* refers to the combination of edge dislocation and screw dislocation.

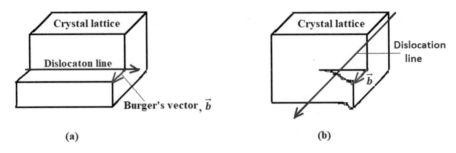

Fig. 2.8 Two types of dislocations; (**a**) edge dislocation, (**b**) screw dislocation

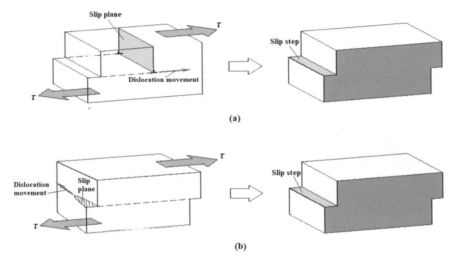

Fig. 2.9 Plastic deformation by slip resulting from the movement of: edge dislocation (**a**), screw dislocation (**b**)

The movement of dislocation results in deformation. There are two principal mechanisms of deformation: slip and twinning; these mechanisms are explained in the following section.

2.4   Deformation Mechanisms – *Dislocation Movement*

2.4.1   *Deformation by Slip*

Slip Planes and Directions   When a shear stress (τ) is applied to a crystalline solid, individual atoms move in a direction parallel to Burger's vector (slip direction) (see Figs. 2.8 and 2.9). On increasing the shear force, the atoms continue to slip to the

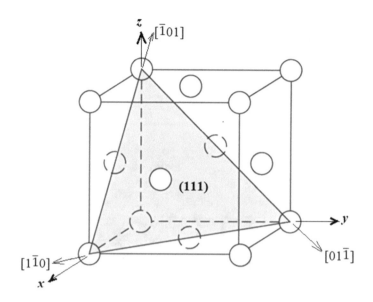

Fig. 2.10   The (111)<110 > slip system in FCC lattice

right (Fig. 2.9). As a row of the atoms find their way back into a proper spot in the lattice, another row of the atoms slip out of position forming a slip step (Fig. 2.9). The process by which plastic deformation results by dislocation movements, is called *slip*; which occurs along specified *slip planes*. A *slip plane* is the crystallo-graphic plane (with the greatest atomic packing) along which the dislocation line traverses under the action of a shear stress. In an edge dislocation, the dislocation moves (in its slip plane) along the slip direction (Fig 2.9a). In a screw dislocation, the dislocation moves (in its slip plane) in a direction perpendicular to the slip direc-tion (Fig. 2.9b). Dislocation movement, resulting in deformation, is also illustrated in Fig. 4.1 (Chap. 4).

Slip Systems   We have learnt in the preceding section that plastic deformation of a crystalline solid is caused by a mechanism called *slip*; which occurs across *slip planes*. A *slip plane* is a crystallographic plane with the closest atomic packing (see Figs. 2.9 and 2.10). The slip across a slip plane occurs in specified directions, called *slip directions;* which is the direction of the maximum atomic density. A combina-tion of *slip planes* and *slip directions* is called a *slip system*; the determination of slip system for a crystal structure is explained in the following paragraph.

It is evident in Fig. 2.10 that the (111) slip plane has three slip directions; these slip directions can be denoted by <110>. It can be also observed in Fig. 2.10 that the FCC unit cell has four slip planes which can be represented by {111}. Thus for the FCC lattice, there are 12 slip system (4 slip planes × 3 slip directions = 12 slip sys-tems); these 12 slip systems can be represented as {111}<110 > (see Table 2.2).

Table 2.2 Slip systems in FCC, BCC, and HCP crystals

Crystal	Slip plane	Slip direction	No. of slip planes	No. of slip directions per plane	No. of slip Systems	Slip systems
FCC	{111}	<110>	4	3	12	{111}<110>
BCC	{110}	<111>	6	2	12	{110}<111>
HCP	{0001}	[1000]	1	3	3	{0001} [1000]

Fig. 2.11 Schematic illustration of deformation by twinning

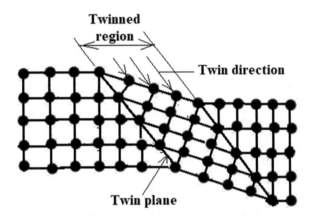

Similarly, for BCC, the number of slip systems can be determined as follows: 6 slip planes × 2 slip directions = 12; these 12 slip systems can be represented by {110} < 111>. Table 2.2 lists the slip systems for BCC, FCC, and CPH crystal structures.

It can be seen in Table 2.2 that each of the two BCC and FCC metals has 12 slip systems; which is much higher as compared to the slip systems in HCP metals. This crystallographic feature justifies an excellent formability exhibited by FCC and BCC metals, in contrast to a poor formability of HCP metals. This mechanical behavior is observed in industrial practice, as follows. Aluminum (FCC) and mild steel (BCC) can be easily plastically deformed by slip to desired shape owing to their good formability, but zinc (HCP) is difficult to be deformed by slip due to its poor formability, owing to a few number (3) of available slip systems (see Table 2.2).

2.4.2 Deformation by Twinning

Twinning is a deformation mechanism that occurs along a definite crystallographic plane in a specific direction which depends on the crystal structure of the metal (see Fig. 2.11). *Deformation by twinning* is different from *deformation by slip* in that the latter can occur along many crystallographic (slip) planes. For examples, in BCC

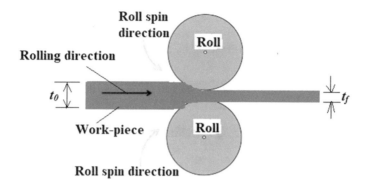

Fig. 2.12 Metal rolling

metals deformation by slip can occur along 6 slip planes with 2 direction on each slip plane; but deformation by twining occurs along just a twin plane (112) in the direction [111]. In HCP metals, the twin plane and direction are (10$\bar{1}$2) and [10$\bar{1}\bar{1}$], respectively.

2.5 Plastic Deformation – Cold Working/Rolling

We have learnt in the preceding section that plastic deformation of a crystalline material may results by *slip*. In the early stages of plastic deformation, slip is essentially on *slip planes*; as deformation proceeds, cross slip takes place on other planes. Thus, the cold-worked structure forms high dislocation density regions that soon develop into networks of intermingled dislocations (see Fig. 4.5, Chap. 4).

All metal-forming manufacturing processes involve plastic deformation. In industrial practice, most metals are worked at ambient temperature (cold worked) by rolling. Metal rolling is a process by which the stock (work-piece) is introduced between a pair of rolls, and then compressed and squeezed. Figure 2.12 illustrates rolling of a metal strip; which is reduced in thickness from t_0 to t_f with width of the strip assumed to be constant during rolling.

True strain in the work (strip) is given by:

$$\varepsilon_{true} = \ln\left(\frac{A_0}{A}\right) = \ln\left(\frac{w*t_0}{w*t_f}\right) = \ln\left(\frac{t_0}{t_f}\right) \tag{2.4}$$

where ε_{true} is the true strain (see Sect. 6.1, Chap. 6); A_0 is the original cross-sectional area; A is the cross-sectional area at any instant during deformation; t_0 is the original thickness; and t_f is the final thickness (see Example 2.8). The amount of thickness reduction is expressed as draft, d, as follows:

$$d = t_0 - t_f \tag{2.5}$$

The amount of deformation is measured by percent cold work (% CW), as follows:

$$\%CW = \frac{A_0 - A_{cw}}{A_0} \times 100 = \frac{t_0 - t_f}{t_0} \times 100 \tag{2.6}$$

Equation 2.6 holds good for flat rolling (rolling of slabs/plates/sheets) (Example 2.9).

2.6   Deformation in Single Crystals – *Schmid's Law*

We learnt in Sect. 2.4 that plastic deformation by slip occurs as a result of disloca-tion motion across slip planes when shear stresses are applied along a slip plane in a slip direction. In case, the applied stress is normal (tensile or compressive), there also occurs deformation by slip; in this case, there exists a shear-stress component in the slip direction inclined at an angle to the applied stress direction. This shear component is called the *resolved shear stress, τ_r* (Gottstein 2004). In order to resolve the applied stress onto the slip plane, we consider a load F that is applied along the tensile axis of a single crystal (Fig. 2.13). This force per unit area is the tensile stress, σ. Now we resolve the tensile force into shear force (F_s) onto the slip plane, as shown in Fig. 2.13.

It is evident in Fig. 2.13 that the tensile force (F) is related to tensile stress (σ) by:

$$F = \sigma A_0 \tag{2.7}$$

Since the shear force F_s acts in the slip direction, we can write:

$$F_s = F \cos \lambda \tag{2.8}$$

Fig. 2.13 Resolution of tensile stress $\left(\sigma = \dfrac{F}{A_0} \right)$ into shear stress $\left(\tau_r = \dfrac{F_s}{A} \right)$ under tensile loading

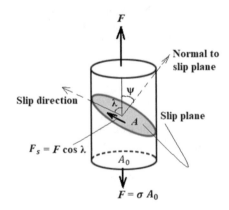

where λ is the angle between the tensile axis and the slip direction.

By combining Eqs. 2.7 and 2.8, we obtain:

$$F_s = \sigma A_0 \cos \lambda \qquad (2.9)$$

By reference to Fig. 2.13, we obtain:

$$A_0 = A \cos \psi \qquad (2.10)$$

By combining Eqs. 2.9 and 2.10, we obtain:

$$F_s = \sigma A (\cos \psi)(\cos \lambda) \qquad (2.11)$$

$$\frac{F_s}{A} = \sigma (\cos \psi)(\cos \lambda) \qquad (2.12)$$

But the shear force (F_s) divided by the slip-plane area (A) is the resolved shear stress (τ_r). Thus,

$$\tau_r = \sigma (\cos \psi)(\cos \lambda) \qquad (2.13)$$

where σ is the tensile stress, ψ is the angle between the tensile axis and normal to slip plane; and λ is the angle between the tensile axis and the slip direction (see Example 2.10).

A single crystal has a number of different slip systems that are capable to cause plastic deformation. Although the resolved shear stress is normally different for each slip system, there is one slip system that is the most favorably oriented to initiate slip. The minimum shear stress required to initiate slip in a single crystal under an applied normal stress, is called the *critical resolved shear stress*, τ_{crss}. The *critical resolved shear stress* refers to the maximum value of the resolved shear stress that results in plastic deformation of the single crystal by yielding. Hence, Eq. 2.13 can be re-written as:

$$\tau_{crss} = \sigma_{ys} \left[(\cos \psi)(\cos \lambda)\right]_{max} \qquad (2.14)$$

where σ_{ys} is the yield strength of the single crystal. Eq. 2.14 is known as **Schmid's law**. The significance of Schmid's law is illustrated in Example 2.11. It has been shown that for single crystals of several metals, the critical resolved shear stress (τ_{crss}) is related to the dislocation density (ρ_D) by (Huda 2020):

$$\tau_{crss} = \tau_0 + A \sqrt{\rho_D} \qquad (2.15)$$

where τ_0 and A are constants; and ρ_D is the dislocation density which is defined as the number of dislocation lines that thread a unit area of surface, # mm^{-2} (see Example 2.12).

2.7 Calculations – *Worked Examples*

Example 2.1 Calculating the Atomic Packing Factor Calculate the atomic packing factor (*APF*) for: (a) BCC, and (b) FCC crystal structures.

Solution (a) No. of atoms per unit cell for BCC = N_a = 2

By reference to Fig. 2.3(b), we get: $R = \dfrac{\sqrt{3}}{4}a$

$$\text{Volume of an atom in } BCC = V_a = \frac{4}{3}\pi R^3 = \frac{4}{3}\pi\left(\frac{\sqrt{3}}{4}a\right)^3 = \frac{\sqrt{3}}{16}\pi a^3 \qquad \text{(E.2.1a)}$$

The volume of BCC unit cell = $V_{uc} = a^3$ (E.2.1b) (see Fig. 2.3a)

By substituting the values from Eqs. E.2.1a – E.2.1b into Eq. 2.1, we get:

$$\text{Atomic packing factor for } BCC \text{ crystal structure} = APF = \frac{N_a V_a}{V_{uc}} = \frac{2\times\dfrac{\sqrt{3}}{16}\pi a^3}{a^3} = 0.68$$

(b) No. of atoms per unit cell for FCC = N_a = 4, Volume of FCC unit cell = $V_{uc} = a^3$

By reference to Fig. 2.4(b), we get: $R = \dfrac{\sqrt{2}a}{4}$

$$\text{Volume of an atom in } FCC = V_u = \frac{4}{3}\pi R^3 = \frac{4}{3}\pi\left(\frac{\sqrt{2}a}{4}\right)^3 - \frac{4}{3}\pi\left(\frac{\sqrt{2}a^3}{32}\right) = 0.185\,a^3$$

By using Eq. 2.1,

$$\text{Atomic packing factor for } FCC \text{ crystal structure} = APF = \frac{N_a V_a}{V_{uc}} = \frac{4\times 0.185\,a^3}{a^3} = 0.74$$

Example 2.2 Calculating the Number of Atoms Per Unit Cell Calculate the
number of atoms per unit cell for: (a) FCC, and (b) HCP crystal structures.

Solution (a) For FCC unit cell, there are 8 one-eighths atoms at the corners and 6
halve atoms at the faces.

Contribution of 8 one – eighth corner atoms to the unit cell $= 8\times\dfrac{1}{8} = 1$ (see Fig. 2.4b)

Contribution of 6 halve atoms at the faces of the unit cell $= 6\times\dfrac{1}{2} = 3$ (see Fig. 2.4b)

Hence, the number of atoms per unit cell in FCC lattice = $1 + 3 = 4$

(b) In the HCP unit cell, each corner atoms contributes one-sixth to the unit cell, and each atom

at the face is shared with the adjacent cell; the 3 middle-layer atoms contribute full to the cell

(see Fig. 2.5).

$$\text{Contribution of 6 corner atoms in the top layer of the unit cell} = 6 \times \frac{1}{6} = 1$$

$$\text{Contribution of 2 face atoms} \left(\text{one in the top layer and the other in the bottom} \right) = 2 \times \frac{1}{2} = 1$$

$$\text{Contribution of 6 corner atoms in the bottom layer of the unit cell} = 6 \times \frac{1}{6} = 1$$

Full contribution of the 3 middle-layer atoms = 3

Hence, the number of atoms per unit cell in HCP lattice = $1 + 1 + 1 + 3 = 6$

Example 2.3 Computing the Volume of HCP Unit Cell Calculate the volume of unit cell in HCP lattice in term of the lattice parameter.

Solution In order to determine the area of the hexagonal face of the HCP unit cell, we can consider a new figure comprising of six triangular divisions of the hexagonal face. Accordingly,

$$\text{Area of hexagonal face} = 6 \times \left(\text{area of each triangle} \right) = 6 \times \left(\frac{1}{2} \times \text{base} \times h \right) = 3a(h) = 3\frac{\sqrt{3}}{2}a^2$$

By reference to Fig. 2.5(a), the volume of HCP unit cell (V_{uc}) can be expressed as:

$$V_{uc} = \left(\text{area of the hexagonal face} \right) \times \left(\text{height of the hexagonal face} \right)$$

$$V_{uc} = \left(3\frac{\sqrt{3}}{2}a^2 \right) \times (c) = \left(3\frac{\sqrt{3}}{2}a^2 \right) \times (1.63a) = 4.23a^3 \qquad \text{(or)}$$

The volume of HCP unit cell = $4.23\,a^3$

Example 2.4 Computing the Atomic Packing Factor for HCP Crystal By using the data in Example 2.3, calculate the APF for HCP Crystal structure.

Solution No. of atoms per unit cell for HCP = N_a = 6; Volume of HCP unit cell = $V_{uc} = 4.23\,a^3$

Fig. 2.14 A direction
vector in a crystal lattice

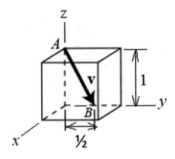

By reference to Fig. 2.5, volume of an atoms (V_a) is related to the HCP lattice
parameter by: $V_a = 0.522\,a^3$

By using Eq. 2.1,

$$APF = \frac{N_a V_a}{V_{uc}} = \frac{6 \times 0.522\,a^3}{4.23\,a^3} = 0.74$$

The atomic packing factor for HCP crystal structure = 0.74

*Example 2.5 Determining the Miller Indices for a Crystallographic
Direction* Define the crystallographic direction for the vector $\bar{\mathbf{v}}$ (in Fig. 2.14)
using Miller Indices.

Solution By reference to Sect. 2.2.5, we find the Miler Indices as follows:

(I)   locating two arbitrary points on direction $\bar{\mathbf{v}}$: A and B,
(II) the coordinates of the two points are: A(0,0,1) and B(0,½ ,0),
(III) performing subtraction: B(0, ½, 0) − A(0, 0, 1) = (0, ½, -1),
(IV) clearing the fraction: 2 (0, ½,-1) = (0,1,-2), and obtaining Miller indices:
(V)   Miller Indices are: $\left[0,1,\bar{2}\right]$.

**Example 2.6 Finding Miller Indices using the shown Crystallographic
Planes** Determine the *Miller indices* for the crystallographic planes shown in
Figs. 2.15(a–c).

Solution By following the step-by-step procedure outlined in Sect. 2.2.6, we get:

Fig. 2.15(a)

	x	y	z
I. Intercepts:	−1	∞	∞
II. Reciprocals:	−1	1/∞	1/∞
III. Miller Indices:		$(\bar{1}00)$	

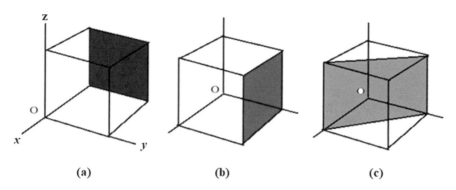

Figs. 2.15 Crystallographic planes with unknown Miller Indices

Fig. 2.15(b)

	x	y	z
I. Intercepts:	∞	1	∞
II. Reciprocals:	1/∞	1/1	1/∞
III. Miller Indices:		(010)	

Fig. 2.15(c)

	x	y	z
I. Intercepts:	1	1	∞
II. Reciprocals:	1/1	1/1	1/∞
III. Miller Indices:		(110)	

Example 2.7 Sketching Crystallographic Planes using the Known Miller Indices Sketch the crystallographic planes with the following Miller Indices: (001), (110), and (011).

Solution   (001) Plane

	x	y	z
Reduction	0	0	1
Reciprocals	1/0	1/0	1/1
Intercepts	∞	∞	1

The (011) plane is sketched in Fig. 2.7 (left).
(110) Plane

	x	y	z
Reduction	1	1	0
Reciprocals	1/1	1/1	1/0
Intercepts	1	1	∞

The (110) plane is sketched in Fig. 2.7 (center).

(011) Plane

	x	y	z
Reduction	0	1	1
Reciprocals	1/0	1/1	1/1
Intercepts	∞	1	1

The (011) plane is sketched in Fig. 2.7 (right).

Example 2.8 Computing the True Strain in Metal Rolling A 50-mm thick brass plate was rolled to 40-mm thickness and then again rolled down to 25 mm thickness. Calculate the true strain in the rolling operation.

Solution $t_0 = 50$ mm, $t_f = 25$ mm, $\varepsilon = ?$

By using Eq. 2.4,

$$\text{True strain in the rolling operation} = \varepsilon_{true} = \ln\left(\frac{t_0}{t_f}\right) = \ln\left(\frac{50}{25}\right) = \ln 2 = 0.693$$

Example 2.9 Computing the Draft and % Cold Work in Rolling By using the data in Example 2.8, calculate the % cold work in the rolling operation.

Solution By using Eq. 2.6,

$$\%CW = \frac{t_0 - t_f}{t_0} \times 100 = \frac{50 - 25}{50} \times 100 = 50$$

Example 2.10 Calculating the Resolved Shear Stress in a Single Crystal A single crystal is acted upon by a tensile stress of 2 MPa thereby causing deformation by slip. The angle between the tensile axis and normal to the slip plane is 60°. The angle between the tensile axis and the slip direction is 50°. Calculate the resolved shear stress along the slip plane.

Solution $\sigma = 2$ MPa, $\psi = 60°$, $\lambda = 50°$, $\tau_r = ?$

By using Eq. 2.13,
Resolved shear stress $= \tau_r = \sigma\ (\cos\psi)\ (\cos\lambda) = 2 \times \cos 60° \times \cos 50° = 0.642$ Mpa

Example 2.11 Calculating the Yield Strength of a Single Crystal by Inputting τ_{crss} A metal single crystal is oriented such that the angle between the tensile axis and normal to slip plane is 60°. Two possible slip directions make angles of 35° and 40° with the tensile axis. The *critical resolved shear stress* for the metal is 6 MPa. Calculate the yield strength of the metal.

Solution $\tau_{crss} = 6$ MPa, $\psi = 60°$, $\lambda_1 = 35°$, $\lambda_2 = 40°$ $\sigma_y = ?$

By using $\lambda_1 = 35°$, $(\cos \psi)(\cos\lambda) = (\cos 60°) \times (\cos 35°) = 0.5 \times 0.82 = 0.41$
By using $\lambda_2 = 40°$, $(\cos \psi)(\cos\lambda) = (\cos 60°) \times (\cos 40°) = 0.5 \times 0.77 = 0.38$
Hence, $[(\cos \psi)(\cos\lambda)]_{max} = 0.41$
By using the modified form of Eq. 2.14,

$$\text{The yield strength of the metal} = \sigma_y = \frac{\tau_{crss}}{\left[(\cos\psi)(\cos\lambda)\right]_{max}} = \frac{6}{0.41} = 14.6\,MPa$$

Example 2.12 Calculating the τ_{crss} by Inputting the Dislocation Density At a dislocation density of 10^5 # mm^{-2}, the critical resolved shear stress (τ_{crss}) of copper is 2.10 MPa. At a dislocation density of 10^4 # mm^{-2}, the τ_{crss} of copper is 0.7 MPa. Calculate the critical resolved shear stress of copper at a dislocation density of 10^6 # mm^{-2}.

Solution By using Eq. 2.15 for $\tau_{crss} = 2.10$ MPa, and $\rho_D = 10^5$ # mm^{-2}

$$\tau_{crss} = \tau_0 + A\sqrt{\rho_D}$$

$$2.10 = \tau_0 + A\sqrt{10^5} \tag{E2.12-A}$$

By using Eq. 2.15 for $\tau_{crss} = 0.7$ MPa, and $\rho_D = 10^4$ # mm^{-2},

$$0.7 = \tau_0 + A\sqrt{10^4} \tag{E2.12-B}$$

By solving Eqs. E2.12-A and E2.12-B simultaneously, we obtain:
  $\tau_0 = 0.069$ MPa, and $A = 6.35 \times 10^{-3}$ MPa-mm
  By using Eq. 2.15 for $\rho_D = 10^6$ # mm^{-2},

$$\tau_{crss} = \tau_0 + A\sqrt{\rho_D} = \left[0.069 + \left(6.35 \times 10^{-3}\right)\sqrt{10^6}\right] = 0.069 + \left(6.35 \times 10^{-3} \times 10^3\right) = 6.42\,MPa$$

At a dislocation density of 10^6 # mm^{-2} in copper, the critical resolved shear stress = 6.42 MPa.

Example 2.13 Calculating the Lattice Parameter for a Metal's Unit Cell The atomic radius of gold (FCC) is 0.1442 nm. Calculate the lattice parameter for gold.

Solution By using Eq. 2.3,

$$\text{The lattice parameter for gold} = a = 2.828R = 2.828 \times 0.1442 = 0.4078\,nm$$

Questions and Problems

2.1. Diagrammatically illustrate the seven crystal systems.

2.2. Explain the applications of crystallographic directions and planes in plastic deformation.

2.3. Why is the (actual) strength of a real crystalline material much lower than the theoretically predicted strength of otherwise perfect crystal?

2.4. (a) Draw sketches for edge dislocation and screw dislocation indicating Burger's vector for each type of dislocation. (b) Explain the role of dislocation movement in deformation.

2.5. Explain deformation by slip with the aid of diagrams.

2.6. (a) Define the following terms: (i) slip plane, (ii) slip direction, (iii) slip system. (b) Draw a sketch showing slip plane and slip directions in a BCC unit cell.

2.7. FCC and BCC metals be easily deformed by slip, but HCP metals not. Explain.

2.8. Differentiate between the following terms: (a) slip and twinning, (b) resolved shear stress and the critical resolved shear stress, (c) amorphous and crystalline solids.

2.9. The atomic radius of silver (FCC) is 0.1445 nm. Compute the lattice parameter for silver

2.10. Sketch the crystallographic planes with the Miller Indices: (100) and $(0\,\bar{1}\,0)$.

2.11. A tensile stress of 1.8 MPa acts on a single crystals causing deformation by slip. The angle between the tensile axis and normal to the slip plane is 56°. The angle between the tensile axis and slip direction is 48°. Calculate the resolved shear stress on the slip plane.

2.12. Calculate the critical resolved shear stress for a single crystal of copper at a dislocation density of 1000 mm^{-2}. Hint: use the data in Example 2.12.

2.13. In a stressed single crystal of zinc, the angle between the tensile axis and normal to the slip plane is 65°. Three possible slip directions make angles of 70°, 50°, and 30° with the tensile axis. Calculate the yield strength of zinc, if its critical resolved shear stress is 0.91 MPa.

2.14. Sketch the crystallographic direction with Miller Indices $[01\,\bar{3}\,]$ and [112].

2.15. A metal plate with thickness 47 mm was rolled to 30 mm thickness and then again rolled down to 12 mm thickness. Calculate the (a) true strain (b) draft, and (c) % reduction.

References

Callister WD (2007) Materials science and engineering: *an introduction*. John Wiley & Sons Inc., Milton

Douglas B, Ho S-H (2006) Structure and chemistry of crystalline solids. Springer, New York

Gottstein G (2004) Physical foundations of materials science. Springer Publishing, Berlin

Huda Z, Bulpett R (2010) Materials science and design for engineers. Trans Tech Publications, Zurich-Durnten

Huda Z (2020) Metallurgy for physicists and engineers. CRC Press, Boca Raton

Helliwell JR (2015) Perspectives in crystallography. CRC Press, Boca Raton

Hull D, Bacon DJ (2011) Introduction to dislocations, 5th edn. Elsevier Science Publications Inc., New York

Vainshtein BK, Friedkin VM, Indenbom VL (1995) Structure of crystals. Springer- Verlag, Berlin

Chapter 3
Mechanical Testing and Properties of Materials

3.1 Material Processing and Mechanical Properties

3.1.1 Relationship Between Processing and Properties

The mechanical behavior of a material strongly depends on its mechanical properties; which in turn also depends on the processing given to the material. For example, the tensile and fatigue strengths of forged (metal formed) components are significantly superior to those of the components manufactured by metal casting. Another example of the dependence of mechanical properties on processing is the effect of rate of cooling (solidification) in metal casting/welding. Rapid cooling results in fine-grained microstructure associated with higher strength; whereas solidification with a slow cooling rate produces coarse-grained microstructure indicating a lower-strength material.

3.1.2 Mechanical Properties/Behaviors

Mechanical properties are the physical properties that a material exhibits in response to an applied load or stress. These properties determine the amounts of deformation which a stressed material can withstand without failure. Mechanical properties and behaviors include: hardness, elasticity, stiffness, plasticity, strength, ductility, toughness, and the like (see Fig. 3.1).

It is evident in Fig. 3.1 that the mechanical properties/behaviors of metals can be divided into four groups: (a) elastic properties, (b) plastic properties, (c) fatigue behavior, and (d) creep behavior. *Elasticity* refers to the ability of a material to be deformed under load and then return to its original shape and dimensions when the load is removed. *Plasticity* is the ability of a material to be permanently deformed

Z. Huda, *Mechanical Behavior of Materials*, Mechanical Engineering Series, https://doi.org/10.1007/978-3-030-84927-6_3

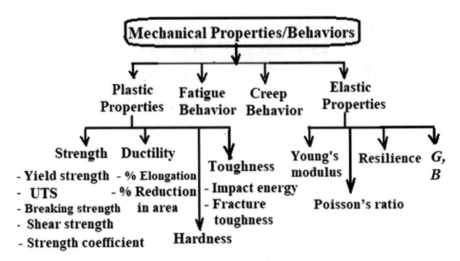

Fig. 3.1 Classification chart of mechanical properties/behavior of materials (*UTS* ultimate tensile strength, *G* shear modulus, *B* bulk modulus)

by applying a load. *Strength* determines the ability of a material to withstand load or stress without failure. *Tensile strength* of a material may be defined as the maximum force required to fracture in tension a bar of unit cross-sectional area. *Hardness* is the ability of a material to resist indentation, scratching, or wear. *Ductility* refers to the ability of a material to undergo deformation under tension without rupture. *Toughness* is the ability of a material to absorb energy when receiving a blow or a shock. These mechanical properties and behaviors are discussed in subsequent sections. The *strength coefficient* is explained in Chap. 7. Additionally, resistance to fatigue and creep are also important mechanical behaviors of engineering materials; these mechanical behaviors are discussed in Chaps. 13 and 14. Another important plasticity behavior is the shear modulus (G); which is explained in the following section.

3.2 Shear Stress and Shear Modulus

When a shear force (F_s) is applied to a solid, an angular or shear deformation is produced in the material (see Fig. 3.2). Thus, a shear stress (τ) produces a shear strain (γ).

By knowing the magnitudes of the shear force and the area over which the force acts, the shear stress can be computed by:

$$\tau = \frac{F_s}{A_s}$$

(3.1)

where F_s is the shear force, N; A_s is the area over which the shear force acts, mm^2; and the τ is the shear stress, MPa (see Example 3.1).

The shear strain (γ) can be determined by reference to Fig. 3.2, as follows:

$$\gamma = \tan\theta = \frac{x}{y}$$

(3.2)

where θ is the angle of twist; x is the transverse displacement, mm; and y is the original length, mm (see Example 3.2). It has been shown that shear stress is directly related to shear strain by:

$$\tau = G\gamma$$

(3.3)

where τ is the shear stress or shear strength, MPa; γ is the shear strain; and G is the shear modulus or the modulus of rigidity, MPa. Eq. 3.3 enables us to define the shear modulus as follows: "*the shear modulus is the ratio of shear stress to shear strain*". A large shear modulus value indicates that the solid is highly rigid *i.e.* a large force is required to produce deformation. On the other hand, a small shear modulus value indicates a soft or flexible solid *i.e.* a little force is needed to deform it. Table 3.1 lists the values of some commonly used engineering materials at ambient temperature. By combining Eqs. 3.1, 3.2, and 3.3, the modulus of rigidity (G) can be calculated by:

$$G = \frac{F_s\, y}{A_s\, x}$$

(3.4)

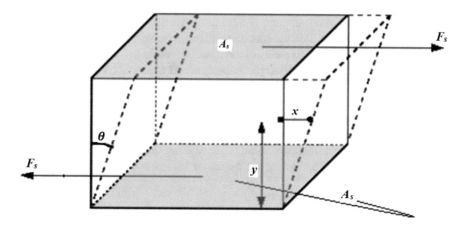

Fig. 3.2 Shear deformation in a solid

Table 3.1 Shear Modulus and the Young's modulus of some materials at ambient temperature

Material	Al_2O_3	PE	Lead	Aluminum	Copper	Nickel	MS	Magnesium
Young's modulus, GPa[a]	385	0.3	15	70	110	207	207	45
Shear modulus, GPa	150	0.22	6	25	46	76	83	17

[a]1 GPa = 10^9 N/m²
PE polyethylene, *MS* mild steel

The significance of Eq. 3.4 is illustrated in Example 3.3. Besides Eq. 3.4, the modulus of rigidity can also be determined by measuring the slope of the linear elastic region in the shear stress-strain curve (Callister 2007).

3.3   Tensile Testing and Tensile Properties

3.3.1   Tensile Testing

A tensile test (or a tension test) is the most fundamental test to determine the tensile mechanical properties of a material. Prior to tensile testing, it is necessary to produce a tensile test specimen, as shown in Fig. 3.3. The dimensions of the test specimen must be in accordance with the standards prescribed by the American Society of Testing of Materials (ASTM). The original gage length (L_o) and the original diameter (d_0) are measured and recorded (see Fig. 3.3a).

The tensile test is conducted by use of a tensile testing machine or universal testing machine; the latter can be used for both tension and compression tests. In tensile testing, the specimen is firmly held in place by grips of the testing machine (see Fig. 3.4). One end of the specimen is held firm, whilst the other end is pulled by applying a tensile force within the specimen. Modern universal testing machines are fitted with instrumentations to measure the force as well as the velocity of the movable crosshead. An electronic device (extensometer) is mounted on the specimen for measuring the specimen extension; the extensometer must be removed once the specimen approaches its proportional limit else it will be damaged when the specimen breaks. Once the material has fractured, the specimen is re-assembled; and the gage length at failure (L_f), and the final cross sectional area (A_f) are measured.

By using the force and dimensional changes data, the engineering stress (σ_{eng}) and engineering strain (ε_{eng}), can be calculated by using the following mathematical relationships:

$$\sigma_{eng} = \frac{F}{A_0}$$

$$(3.5)$$

Fig. 3.3 Tensile test
specimen before loading
(**a**), and after loading (**b**)

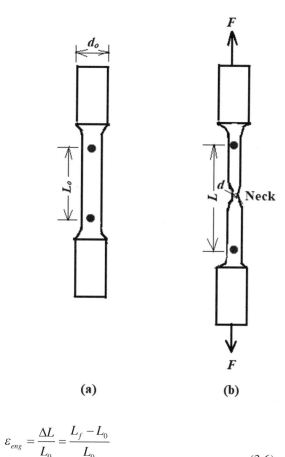

(**a**)   (**b**)

$$\varepsilon_{eng} = \frac{\Delta L}{L_0} = \frac{L_f - L_0}{L_0}$$

$$(3.6)$$

where σ_{eng} is the engineering stress, MPa; F is the applied force, N; A_o is the original cross-sectional area of the specimen, mm^2; ε_{eng} is the engineering strain; L_o is the original length of the specimen (or distance between gage marks), mm; and L_f is the final gage length just before fracture, mm (see Example 3.4). The force and the dimensional changes data recorded during the tensile test are used to calculate various tensile mechanical properties (see Eqs. 3.7, 3.8, 3.9. 3.10, 3.11, and 3.12).

3.3.2 Tensile Mechanical Properties

Stress-Strain Curve In the preceding sub-section, we learnt that tensile testing results in deformation (elongation) of the specimen until it breaks. A graphical representation of the relationship between load applied and deformation of the material is expressed as a stress-strain curve. Thus, the stress-strain curve presents a complete tensile profile, as shown in Fig. 3.5.

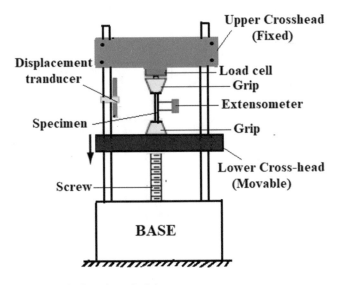

Fig. 3.4 Tensile (mechanical) testing principle

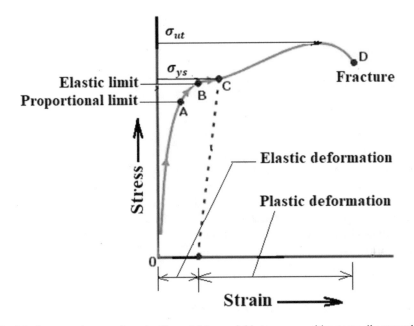

Fig. 3.5 Stress-strain curve for a ductile metal (σ_{ys} = yield stress, σ_{ut} = ultimate tensile strength)

Strength Properties The first stage of the stress-strain curve (Fig. 3.5) indicates elastic behavior of material; here the stress is proportional to the elastic strain up to the proportional limit (at point *A*) *i.e. Hook's Law* is obeyed from *O* to *A*. This linear elastic behavior is expressed as:

$$\sigma = E\varepsilon_e \tag{3.7}$$

where σ is the *stress* in the elastic range; ε_e is the *elastic strain*; and the constant of proportionality E is called *the Young's modulus* or modulus of elasticity. The Young's modulus is a measure of the stiffness of the material in tension.

Beyond the proportional limit, the straight line becomes a curve. The stress at the elastic point B is called the *elastic limit* (see Fig. 3.5). Then, the material starts to yield at the yield point C. The stress-strain relationship for yielding can be expressed by modifying Eq. 3.7 as follows:

$$\sigma_{ys} = E\epsilon_{yield} \tag{3.8}$$

where σ_{ys} is the yield stress or yield strength, MPa; and ϵ_{yield} is the elastic strain taken to yielding.

The modulus of elasticity (E) values of some metals are listed in Table 3.1. The Young's modulus may be computed by re-writing Eq. 3.7 as follows:

$$E = \frac{stres}{Elastic\ strain} = \frac{\dfrac{F}{A_0}}{\dfrac{\delta l}{L_0}} = \frac{F\,L_0}{A_0\,(\delta l)} \tag{3.9}$$

where F is the force up to proportional limit, N; δl is the elastic elongation, mm; and E is the Young's modulus, MPa (see Examples 3.5, 3.6, and 3.7).

Beyond the yield point (C), the material *strain hardens i.e.* there is an increase in stress with increasing strain (see Fig. 3.5; see also Chap. 7). The maximum stress during the tensile test, is called the *ultimate tensile strength* or *tensile strength (σ_{ut})*; which can be computed by:

$$\sigma_{ut} = \frac{F_{max}}{A_0} \tag{3.10}$$

where σ_{ut} is the ultimate tensile strength (or tensile strength), MPa; F_{max} is the maximum force acting on the material during the tensile test, N; and A_0 is the original cross-sectional area, mm². On reaching the maximum stress (σ_{ut}), necking starts and fracture finally occurs at the point D (see Figs. 3.3 and 3.5). The stress at the fracture point, is called the *failure strength* or the *breaking strength*. The significance of Eq. 3.10 and Fig. 3.5 is illustrated in Example 3.8. It is worth noting that from the point B to D, the material is strained by plastic deformation.

Ductility Properties The stress-strain curve also enables us to determine ductility properties. The percent elongation and the percent reduction in area (% RA) can be calculated as follows.

$$\%\text{Elongation} = \frac{\Delta L}{L_0} \times 100 = \frac{L_f - L_0}{L_0} \times 100$$

(3.11)

$$\%\text{Reduction in Area} = \%\text{RA} = \frac{A_0 - A_f}{A_0} \times 100$$

(3.12)

The significance of Eq. 3.11 is illustrated in Example 3.9. The tensile mechanical properties of some engineering materials are presented in Table 3.2.

The stress-strain curve, as obtained from the tensile test of a mild steel (AISI-1020) sample, is shown in Fig. 3.12 (see also Chap. 5, Section 5.5.1 for the AISI system of steel). The stress-strain curve (Fig. 3.12) enables us to determine the tensile mechanical properties of AISI-1020 steel (see Examples 3.7, 3.8, 3.9, and 3.10).

3.4   Elastic Mechanical Properties

In the preceding section, we learnt that the Young's modulus (E) is an elastic constant of a material that determines its stiffness in tension. Another elastic constant is the shear modulus (G); which is the ratio of shear stress to the shear strain (see Sect. 3.2). Besides E and G, there is yet another elastic constant – the bulk modulus (B); which is explained in Chap. 7 (Sect. 7.5). Additionally, there are two other important elastic properties: *Poisson's ratio* (ν) and *resilience* (U_r); these elastic properties are discussed in Chap. 7. It is worth mentioning that the Young's modulus of many materials strongly depend on their crystallographic/phase orientations; such materials are called *anisotropic or orthotropic materials*. It means that anisotropic materials have different properties in different directions (see Chap. 15, Sect. 15.2). When the mechanical behavior of a material is same for all crystallographic/phase

Table 3.2 Tensile mechanical properties of some materials

Material	Yield strength (σ_{ys}), MPa	Tensile strength (σ_{ut}), MPa	% Elongation (in $L_0 = 50$ mm)
Polyethylene (high density)	26–30	22–30	11.2–13.0
Alumina (Al_2O_3)	n/a	300	0.001–0.01
Aluminum	35	90	40
Copper	70	200	45
Steels	220–1000	350–2000	12–30
Cast irons	100–150	300–1000	0–7
Nickel	140	480	40
Magnesium alloys	170–200	300–320	10–13
Titanium	450	520	25

orientations, the material is called *isotropic*. The elastic behavior of *isotropic materials* is expressed in terms of *Poisson's ratio* (v).

The elastic mechanical properties and behavior of materials is discussed in detail in Chap. 7.

3.5 Hardness Testing and Hardness of Materials

3.5.1 Hardness and its Testing

The ability of a material to resist plastic deformation by indentation (penetration), scratching, or abrasion is referred to as hardness. A hard material has the ability to resist permanent deformation (bending, breaking, or change in its shape), when a load is applied on it. In general, hardness is measured as the indentation hardness; which is defined as the resistance of a material to penetration or indentation. Hardness testing involves the use of a pointed or rounded indenter that is pressed into a surface under a substantially static load (Pelleg 2013). Hardness tests are routinely conducted in manufacturing industries for specification purposes, for checking the effectiveness of surface-hardening methods, for checking heat treatment procedures and/or as a substitute for tensile tests on small parts. There are five commonly-used methods of hardness testing: (a) Brinell hardness test, (b) Rockwell hardness test, (c) Vickers test, (d) Knoop hardness test, and (e) Microhardness test. These testing methods are briefly described in the following sub-sections.

3.5.2 Brinell Hardness Test

In the Brinell hardness test, the test material is indented by use of a 10-mm-diameter hardened- steel/carbide ball that is subjected to a specified load, F (see Fig. 3.6). For hard materials, the full load of 3000 kgf is applied; whereas for softer materials, the applied load is in the range of 500–1500 kgf. The full load is normally applied for 10–15 seconds in the case of hard materials; this duration is about 30 s in the case of softer materials. A low-power microscope is used to measure the diameter of the indentation produced in the test material (see Fig. 3.6b).

Brinell harness number (BHN) can be calculated by dividing the load applied (F) by the surface area of the indentation, according to the following formula:

$$\text{BHN} = \frac{F}{\frac{\pi}{2}\left(D - \sqrt{D^2 - D_i^2}\right)}$$

(3.13)

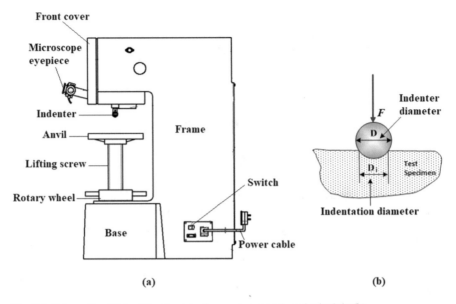

Fig. 3.6 Schematic of Brinell hardness testing equipment (**a**), and principle (**b**)

where D and D_i are the diameters of indenter and indentation, respectively (see Example 3.11).

A well-defined BHN indicates the test conditions. For example, 90 HB 10/1000/20 means that Brinell Hardness of 90 was obtained using a 10-mm-diameter hardened steel ball with a 1000 kgf load applied for 20 seconds. For testing of extremely hard metals, the steel ball is replaced by a tungsten carbide (WC) ball. Brinell hardness test is the most suitable method for achieving the bulk or macro-hardness of a material, particularly those with heterogeneous structures. Table 3.3 lists BHN values of some materials.

For most steels, there exists a relationship between Brinell hardness and strength according to:

$$\sigma_{ut} = 4.35(\text{BHN}) \tag{3.14}$$

where σ_{ut} is the ultimate tensile strength, MPa; and *BHN* is the Brinell hardness number.

3.5.3   Rockwell Hardness Test

Rockwell hardness testing involves indenting the test material by using a hardened steel-ball indenter or a diamond cone-shaped Brale indenter. Initially, the indenter is forced into the test material under a minor load *F0,* usually 10 kgf (see Fig. 3.7). On

Table 3.3 Brinell hardness numbers of some materials at 25 °C

Material	Most plastics	Brasses & Al alloys	Carbon steels	Cast irons	Hardened steels
BHN	8–30	60–100	130–235	350–415	800–900

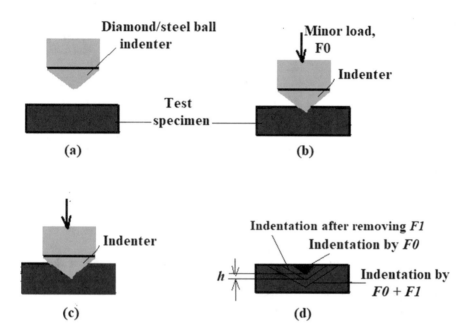

Fig. 3.7 Rockewell hardness testing principle (h = permenant increase in depth of penetration)

reaching equilibrium, an indicating device is set to a datum position. While the minor load is kept applied, an additional major load ($F1$) is applied with resulting increase in penetration. Once equilibrium has again been reached, the major load is removed but the minor load is still kept maintained. The removal of the major load results in the reduction of the depth of penetration h (see Fig. 3.7).

Rockwell hardness number (HR) is calculated by (Huda 2020):

$$HR = R - h \tag{3.15}$$

where h is the permanent increase in the depth of penetration, measured in units of 0.002 mm (see Fig. 3.7); and R is a constant depending on the form of indenter: $R = 100$ units for diamond indenter, and $R = 130$ units for steel ball indenter.

Rockwell hardness testing indicates the final test results directly on a dial which is calibrated with a series of scales; this testing method is employed for rapid testing of finished materials and products in industry. In general, scales A−C are used for metals and alloys (see Table 3.4). Rockwell hardness numbers for steels range from

Table 3.4 Rockwell Hardness Scales and Loads for typical applications

Scale	Indenter	Minor Load $F0$, kgf	Major Load $F1$, kgf	R	Typical applications
A	Diamond cone	10	50	100	Cemented carbides
B	$\frac{1}{16}$" steel ball	10	90	130	Copper alloys, mild steel, aluminum alloys, malleable irons
C	Diamond cone	10	140	100	Steels, hard cast irons, case hardened steel and other materials harder than 100 HRB

20 HRC (for mild steel) to 80 HRC (for nitrided steels). Brasses and aluminum alloys have hardness numbers in the range of 20 HRB – 60 HRB. This author, in an industrial R&D project, has practically achieved the hardness of 50 HRC in a forged-hardened-tempered steel for cars' axle-hubs (Huda 2012).

3.5.4   Vickers Hardness Test

Vickers hardness testing involves indenting the test material with a diamond indenter. The indenter is in the form of a right pyramid with a square base; the angle between the opposite faces is 136° (see Fig. 3.8a). The indenter is subjected to a load in the range of 1−100 kgf. The full load is generally applied for 10–15 seconds. The two diagonals of the indentation on the surface of the material after removal of the load are measured using a microscope (see Fig. 3.8b).

The indentation diagonals d_1 and d_2 (in mm) enable us to calculate d; which is the arithmetic mean of the two diagonals (see Fig. 3.8b). Vickers hardness number (VHN) is calculated by:

$$\text{VHN} = \frac{2F\sin\frac{136}{2}}{d^2} = \frac{1.854\,F}{d^2}$$

(3.16)

where F is the force applied to the indenter, kgf (see Example 3.12).

Vickers hardness number should be reported in the form: 800 HV/60, which means a Vickers hardness of 800, was obtained using a 60 kgf force. The advantages of Vickers hardness test include: (a) high accuracy in readings, (b) the use of just one type of indenter for all types of metals/alloys and surface treated materials, and the like.

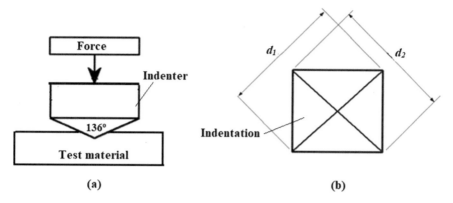

Fig. 3.8 Vickers hardness testing principle (**a**), and the indentation diagonals (**b**)

3.5.5   Knoop Hardness Test

Knoop hardness test, or Knoop microhardness test, involves the use of a rhombic-based pyramidal diamond indenter that forms an elongated diamond shaped indentation. The diamond indenter is pressed into the test material by applying an accurately controlled force, P, in the range of 1–1000 grams. The load is maintained for a specific dwell time: 10–15 s. Then, the indenter is removed leaving an elongated diamond shaped indentation in the test material. An optical microscope is used to measure the longest diagonal (L) of the diamond shaped indentation (see Fig. 3.9).

Knoop hardness number (KHN) is calculated by using the following formula:

$$\text{KHN} = \frac{14.23\,P}{L^2}$$

(3.17)

where P is test force, kgf; and L is the longest diagonal of indentation, mm (see Example 3.13).

The Knoop hardness number normally ranges from HK60 to HK600 for brasses, aluminum alloys, and soft steels. Cutting tools and nitride steels have hardness from HK700 to HK1000.

3.5.6   Microhardness Test

Microhardness testing is based on either Knoop hardness test or Vickers hardness testing. It is used when test samples are very small or thin, or when coating thickness of a plated metal is to be measured. Additionally, small regions in a multiphase metal sample can also be microhardness tested. In the microhardness test, a

Fig. 3.9 Knoop micro-
hardness testing principle

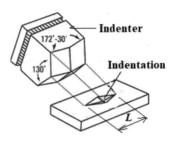

small diamond indenter with pyramidal geometry is forced (1–1000 gf) into the surface of the test material. The resulting indentation is measured under an optical microscope. Eq. 3.16 may be used to calculate VHN, whereas Eq. 3.17 may be used to compute KHN. Modern microhardness testing machines have been automated by linking the indenter apparatus to a computerized image analyzer; this system is capable to determine VHN, KHN, indent location, indent spacing, and the like.

3.5.7   Hardness Conversion

In industrial practice, it is often required to convert hardness on one scale to another. The hardness conversion Tables are available elsewhere (Kinney 1957; Gordonengland 2021). For example, the hardness of a free-cutting steel may be: 200 HB = 90 HRB = 20 HRC = 250 HK.

3.6   Impact Toughness – *Impact Energy*

3.6.1   *Impact Testing*

Impact testing involves the use of a pendulum with a known mass at the end of its arm. The pendulum swings down and strikes a notched specimen which is held securely in position (see Fig. 3.10a). The heavy pendulum, released from a known height, strikes the specimen on its downward swing thereby fracturing it. There are two principal methods for impact testing: (a) Charpy impact test, and (b) Izod impact test. In the *Charpy impact test*, the test specimen is securely held in a horizontal position; and the notch is positioned facing away from the striker (see Fig. 3.10b). In the *Izod impact test*, the notched specimen is held in a vertical position with the notch positioned facing the striker. Impact tests enable us to determine the impact toughness and to assess the notch sensitivity of engineering materials. *Impact toughness* is defined as the amount of energy absorbed before fracturing of the test material when loaded under an impact condition (at a high strain rate).

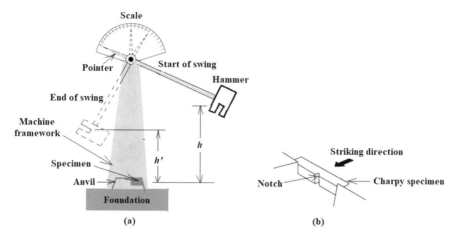

Fig. 3.10 Impact testing principle (**a**), and Charpy test specimen (**b**)

3.6.2 The Analysis of Impact Testing

The impact energy can be calculated by:

$$U_i = m\, g\left(h - h'\right)$$

(3.18)

where U_i is the impact energy of the specimen's material, J; m is the mass of the pendulum, kg; and g is the gravitational acceleration ($g = 9.81$ m/s^2); and $(h - h')$ is the difference between its initial and final heights of the pendulum (see Example 3.14).

The impact toughness of some engineering materials is presented in Table 3.5.

Materials having high impact energies are considered as *ductile materials* (*e.g.* aluminum, lead, mild steel, etc.). On the other hand, materials with low impact energies are *brittle*; notable examples of brittle materials include: glass, cast iron, etc. (see Table 3.4). The impact energy of a metal strongly depends on temperature. In particular, there is a sharp reduction in impact energy when the temperature falls below the ductile-brittle transition temperature (DBTT); which is generally below 0 °C. The ductile-to-brittle transition (DBT) material behavior is crucial to fracture behavior of materials (see Chap. 1).

3.7 Fatigue and Creep Behaviors

In the previous sections, we have learnt about the main mechanical properties of materials; which include the tensile properties, hardness, and impact toughness. However, in industrial practice, there may be special loading conditions (*e.g.* cyclic stresses, high temperature, etc.) to which a material may be subjected during service. Under cyclic stresses, a material may fail by *fatigue*, whereas failure by *creep*

Table 3.5 The impact energies of some materials at 25 °C

Material	PVC	6061 Al alloy	1045 CW steel	1045 normalized steel	1095 normalized steel	304 SS
Charpy impact energy (J)	7	35	15	74	50	130

PVC poly-vinyl chloride, *CW* cold worked, *SS* stainless steel

occurs under long-term loading or loading at a high temperature. Since fatigue and creep reflect the failure mechanical behaviors, they are discussed in Chaps. 13 and 14.

3.8 Calculations – *Worked Examples*

Example 3.1 Computing the Shear Stress in a Solid By using the data in Fig. 3.11, calculate the shear stress acting on the solid.

Solution By reference to Fig. 3.11, $F_s = 10 \times 10^6 N = 10^7 N$, $A_o = 60$ mm $\times$ 25 mm $= 1500$ mm^2

By using Eq. 3.1,

$$\text{The shear stress acting on the solid} = \tau = \frac{F_s}{A_o} = \frac{10^7}{1500} = 6.67 \times 10^3 \, \text{MPa} = 6.67 \, \text{GPa}$$

Example 3.2 Calculating the Shear Strain and the Angle of Twist in a Solid By reference to Fig. 3.11, calculate the (a) shear strain, and (b) angle of twist in the solid.

Solution By reference to Fig. 3.11, $x = 6$ mm, $y = 1.2$ cm $= 12$ mm, $\gamma =?$, $\theta =?$

(a) By using Eq. 3.2,

$$\gamma = \frac{x}{y} = \frac{6}{12} = 0.5$$

(b) Again by using Eq. 3.2,

$$\gamma = \tan \theta = 0.5$$

$$\theta = \tan^{-1} 0.5 = 26.5^\circ$$

The shear strain is 0.5 and the angle of twist is 26.5°.

Example 3.3 Calculating the Shear Modulus for a Solid Material By reference to Fig. 3.11, determine the shear modulus of the solid material by (a) using Equation 3.4, and (b) using the definition of the shear modulus

Solution   By reference to Fig. 3.11, $F_s = 10^7$ N, $A_o = 1500$ mm², $x = 6$ mm, $y = 12$ mm.

(a) By using Eq. 3.4,

$$G = \frac{F_s\, y}{A_s\, x} = \frac{10000000 \times 12}{1500 \times 6} = 13.310^3\,\text{MPa} = 13.3\,\text{GPa}$$

$$G = \frac{\tau}{\gamma}$$

(b) The shear modulus is defined as the ratio of shear stress to shear strain *i.e.*

From Examples 3.1 and 3.2, Shear stress $= \tau = 6.67$ GPa, Shear strain $= \gamma = 0.5$

$$G = \frac{\tau}{\gamma} = \frac{6.67\,GPa}{0.5} = 13.34\,\text{GPa}$$

Example 3.4 Calculating the Engineering Stress and Strain   A tensile force of 1.5 kN is applied to a 130-mm long metal bar with diameter of 2 mm. The final length at failure is 155 mm. Calculate the: (a) engineering stress, (b) engineering strain, and (c) final cross-sectional area.

Solution $L_o = 130$ mm, $L_f = 155$ mm, $d_0 = 2$ mm, F = 1500 N, $\sigma_{eng} = ?$, $\varepsilon_{eng} = ?$, A = ?

$$A_0 = \frac{\pi}{4} d_0^2 = \frac{\pi}{4}(2)^2 = 3.142\,\text{mm}^2$$

(a) By using Eqs. 3.5 And 3.6,

$$\text{Engineering stress} = \sigma_{eng} = \frac{F}{A_0} = \frac{1500}{3.142} = 477.4\,\text{MPa}$$

$$\text{Engineering strain} = \varepsilon_{eng} = \frac{\Delta L}{L_0} = \frac{L_f - L_0}{L_0} = \frac{155 - 130}{130} = 0.19$$

(b)

(c) By using the constant volume relationship,

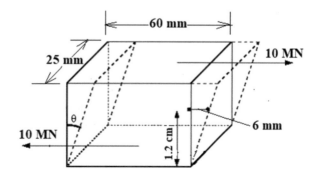

Fig. 3.11 The dimensions of a cubic solid acted upon by a shear force

60 mm

25 mm

10 MN

10 MN

θ

6 mm

1.2 cm

$$A_f L_f = A_0 L_0$$

$$A_f (155) = 3.142 \times 130$$

$$A_f = 2.63 \, \text{mm}^2$$

Example 3.5 Calculating the Young's modulus of a Material, and Identifying it A tensile force of 2.5 kN is applied to a 145-mm-long bar with a diameter of 1.8 mm. The length at the proportional limit is 147 mm. Calculate the Young's modulus of the bar's material; and hence identify the material.

Solution $l_0 = 145$ mm, $d_0 = 1.8$ mm, $l = 147$ mm, $F = 2500$ N, $E = ?$

$$A_0 = \frac{\pi}{4} d_0^2 == \frac{\pi}{4} = 1.8^2 = 2.54 \, \text{mm}^2 \qquad \delta l = l - l_0 = 147 - 145 = 2 \, \text{mm}$$

By using Equation 3.9,

$$\text{Young's modulus} = E = \frac{F l_0}{A_0 (\delta l)} = \frac{2500 \times 145}{2.54 \times 2} = 71{,}358 \, \text{MPa} = 71.36 \, \text{GPa}$$

By reference to Table 3.1, we can identify the material as aluminum.

Example 3.6 Calculating the Tensile Force within the Proportional Limit A tensile force is applied to an aluminum bar of length 170 mm and 2.5 mm diameter. The length of the rod at the proportional limit is 180 mm. Calculate the tensile force.

Solution $l_0 = 170$ mm, $d_0 = 2.5$ mm, $l = 180$ mm, $F = ?$

$$A_0 = \frac{\pi}{4} d_0^2 == \frac{\pi}{4} 2.5^2 = 4.9 \, \text{mm}^2 \qquad \delta l = l - l_0 = 180 - 170 = 10 \, \text{mm}$$

By re-writing Equation 3.9, we obtain:

$$\text{The tensile force} = F = \frac{E A_0 (\delta l)}{l_0} = \frac{69{,}000 \times 4.9 \times 10}{170} = 19.88 \, \text{kN}$$

Example 3.7 Determining the Young's Modulus by using the Stress-Strain Curve Figure 3.12 shows the stress-strain curve for AISI-1020 steel. (a) Identify the proportional limit, and (b) calculate the Young's modulus for AISI-1020 steel.

Solution (a)   By reference to Fig. 3.12, the proportional limit = stress = σ = 370 MPa

(b) The elastic strain corresponding to the proportional limit is 0.0018 (see Fig. 3.12).
   By using the modified form of Eq. 3.7,

$$\text{Young's modulus of AISI   1020 steel} = E = \frac{\sigma}{\varepsilon_e} = \frac{370}{0.0018} = 205.5 \times 10^3 \, \text{MPa} = 205.5 \, \text{GPa}$$

Example 3.8 Determining the Strength Properties from the Stress-Strain Curve By reference to Fig. 3.12, calculate the following strength properties of AISI-1020 steel: (a) tensile strength, (b) yield strength, and (c) fracture strength.

Solution By reference to Fig. 3.12, the maximum stress during the tensile test is σ_{ut} = 505 MPa.

The yield stress = σ_{ys} = 380 MPa. The stress at fracture is 320 MPa.
   Tensile strength = σ_{ut} = 505 MPa, Yield strength = σ_{ys} = 380 MPa, Fracture strength = 320 MPa

Example 3.9 Calculating the Percent Elongation from the Stress-Strain Curve By reference to Fig. 3.12, calculate the percent elongation for the AISI-1020 steel.

Solution By reference to Eqs. 3.6, and 3.11, and Fig. 3.12,

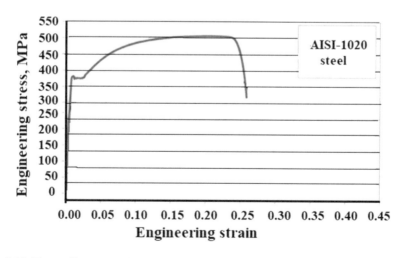

Fig. 3.12 The tensile stress-strain curve for AISI-1020 steel (l_0 = 50 mm)

$$\%\text{Elongation} = \frac{\Delta l}{l_0} \times 100 = \left(\text{Engineering strain at fracture}\right) \times 100 = 0.25 \times 100 = 25$$

Example 3.10 Calculating the Final Length at Fracture in Tensile Test By using the data in *Example 3.9*, calculate the final length at fracture in the tensile test.

Solution By reference to the figure caption of Fig. 3.12, l_0 = 50 mm, % Elongation = 25

By using Eq. 3.11,

$$\%\text{Elongation} = \frac{\Delta l}{l_0} \times 100 = \frac{l_f - l_0}{l_0} \times 100$$

$$25 = \frac{l_f - 50}{50} \times 100$$

$$100 l_f - 5000 = 1250$$

Final length at fracture = l_f = 62.5 mm

Example 3.11 Calculating the Brinell Hardness Number when the Load is Known A load of 1000 kgf is applied by using a 10-mm-diameter steel ball that results in the indentation diameter of 8 mm in a test material. Calculate the BHN for the test material.

Solution F = 1000 kgf, D = 10 mm, D_i − 8 mm, BHN =?

By using Eq. 3.13,

$$\text{BHN} = \frac{F}{\frac{\pi}{2}\left(D - \sqrt{D^2 - D_i^2}\right)} = \frac{1000}{\frac{\pi}{2}\left(10 - \sqrt{10^2 - 8^2}\right)} = \frac{1000}{\frac{\pi}{2}(10 - 6)} = 159$$

Example 3.12 Computing the Vickers hardness number of a test material A load of 80 kgf is applied for 10 s to the indenter of a Vickers hardness tester. The two diagonals of the indentation left on the surface of the test material are 0.8 mm and 1 mm, respectively. Calculate the Vickers hardness number of the test material. Completely specify the hardness.

Solution F = 80 kgf, $d = \dfrac{d_1 + d_2}{2} = \dfrac{0.8 + 1}{2} = 0.09\,\text{mm}$

By using Equation 3.16,

$$\text{VHN} = \frac{1.854\,F}{d^2} = \frac{1.854 \times 80}{0.9^2} = 183$$

The hardness number is 183 VHN/80

Example 3.13 Computing the Knoop hardness number A Knoop indenter presses a test material with a load of 400 gf. The longest diagonal of the indent is measured to be 85 μm. Calculate the Knoop hardness number for the test material.

Solution P = 400 gf = 0.4 kgf, L = 85 μm = 0.085 mm, KHN =?

By using Eq. 3.17,

$$\text{Knoop hardness number} = KHN = \frac{14.23\,P}{L^2} = \frac{14.23 \times 0.4}{0.085^2} = \frac{5.692}{0.00722} = 788$$

Example 3.14 Computing the Impact Energy and Assessing its Ductility/ Brittleness A Charpy impact test was conducted by using a heavy pendulum of 30 kgf that is released from a height of 25 cm. The pendulum struck the specimen on its downward swing thereby fracturing it. The height at the end of swing is 2 cm. Calculate the impact energy of the test material. Is the material brittle or ductile?

Solution m = 30 kgf, g = 9.81 m/s², $h - h' = 25 - 2 = 23$ cm = 0.23 m, U_i = ?

By using Equation 3.18,

$$U_i = m\,g\left(h - h'\right) = 30 \times 9.81 \times 0.23 = 67.7\,J$$

The impact energy of the material is high; hence the material is ductile.

Questions and Problems

3.1 (MCQs). Encircle the most appropriate answer for the following questions.

(a) Which of the following mechanical tests involves the highest strain rate?
  (i) tensile test, (ii) hardness test, (iii) impact test, (iv) shear test
(b) Which of the following mechanical tests involves indentation in the test material?
  (i) hardness test, (ii) shear test, (iii) impact test, (iv) tensile test.
(c) Which of the following mechanical tests is the best for determining the yield strength?
  (i) hardness test, (ii) impact test, (iii) shear test, (iv) tensile test
(d) Which of the following mechanical properties involves elastic energy absorption?
  (i) impact toughness, (ii) resilience, (iii) Young's modulus, (iv) ductility
(e) Which of the following hardness tests involves the use of rhombic-based pyramidal diamond indenter?
  (i) Knoop test, (ii) Brinell test, (iii) Rockwell test, (iv) Vickers test.

(f) Which of the following terms describes the ratio of the lateral strain to the axial strain?

(i) true strain, (ii) engineering strain, (iii) strain ratio, (iv) Poisson's ratio

(g) Which of the following mechanical tests involves the angle of twist?

(i) impact test, (ii) hardness test, (iii) tensile test, (iv) shear test

(h) Which hardness test allows us to use hardened steel ball indenter?

(i) Brinell test, (ii) Vickers test, (iii) Rockwell test, (iv) Knoop test.

3.2 Does there exist a strong relationship between a material's mechanical behavior and its processing? Support your answer by giving at least two example.

3.3 (a) Differentiate between elasticity and plasticity.

(b) Draw the classification chart showing the various mechanical properties of materials.

3.4 (a) Draw the sketch of a tensile test specimen, and label it.

(b) Briefly describe the tensile testing of materials.

(c) List the various strength and ductility properties as obtained from a tensile test.

3.5 Describe the Vickers hardness test with the aid of sketch.

3.6 Which mechanical property must be high in mechanical springs? Explain.

3.7 A tensile force of 2 kN is applied to a 160-mm-long metal bar with a diameter of 2 mm. The final length of the bar is 175 mm. Calculate the: (a) engineering stress, and (b) engineering strain.

3.8 Table 3.6 shows the tensile test data for a ductile material; the original gage length of the tensile-test specimen was 50 mm. (a) Plot the stress-strain curve for the material. (b) Identify the proportional limit and hence calculate the Young's modulus. (c) Calculate the percent elongation for the material. (d) What is the final gage length after fracture?

3.9 By using the data in Table 3.6, determine the tensile strength and the breaking strength for the test material.

3.10 A load of 1000 kgf is applied by using a 10-mm-diameter steel ball that results in the indentation diameter of 9 mm in a test material. Calculate the BHN for the test material.

3.11 The indenter of a Vickers hardness tester is subjected to a load of 70 kgf for 15 s. The two diagonals of the indentation left on the surface of the test material are 0.7 mm and 0.8 mm, respectively. Calculate the Vickers hardness number of the material.

Table 3.6 Tensile Test data for the ductile material

Stress (σ_{eng}), MPa	25	50	75	100	125	150	175	200	225	250	255	255 Break
Strain (ε_{eng}), mm/mm	0.001	0.002	0.003	0.004	0.005	0.01	0.04	0.05	0.07	0.10	0.12	0.15

3.12 A Charpy impact test was conducted by using a heavy pendulum of 2.5 kgf that is released from a height of 15 cm. The pendulum struck the specimen on its downward swing thereby fracturing it. The height at the end of swing is 5 cm. Calculate the impact energy of the test material. Is the material brittle or ductile?

References

Callister WD (2007) Materials science and engineering: an introduction. John Wiley & Sons, Inc., New York

Gordenengland (2021) Hardness conversion table, Internet Source: https://www.gordonengland. co.uk/hardness/hardness_conversion_1c.htm. 27 Feb 2021

Huda Z (2020) Metallurgy for physicists and engineers. CRC Press, Boca Raton, FL

Huda Z (2012) Reengineering of manufacturing process design for quality assurance in axle-hubs of a modern motor-car --- a case study. Int J Automotive Eng 13(7):1113–1118

Kinney GF (1957) Engineering properties and applications of plastics. John Wiley & Sons Inc., New York

Pelleg J (2013) Mechanical properties of materials. Springer Publishing, Netherlands

Chapter 4
Strengthening Mechanisms in Metals/ Alloys

4.1 Strengthening Mechanisms – *Importance and Basis*

Importance of Strengthening Mechanisms The mechanical behavior of metallic materials plays a key role in today's competitive industrial world. Materials technologists often face challenges to design and develop metals and alloys with greater strength for applications in highly-stressed conditions. These challenging technological environments demand controlled strengthening of alloys by using appropriate strengthening mechanisms. In Chap. 2, we have been introduced to the crystalline imperfections with particular reference to dislocations. The movement of dislocation in a crystalline solids is of fundamental importance in understanding strengthening mechanisms in metals (Argon 2007).

The Basis of Strengthening Mechanisms We learnt in Chap. 2 that when a crystal is subjected to shear stress, there are atomic displacements due to dislocation motion (see Sect. 2.4). The basic principle of strengthening mechanism can be illustrated with reference to a crystal lattice containing an edge dislocation. Figure 4.1(a-c) shows a crystal lattice under shear stress (τ). When the resolved shear stress reaches the critical value τ_{crss}, the movement of edge dislocation occurs; which results in plastic deformation. It is evident in Fig. 4.1 that the edge dislocation moves in the direction of applied stress (from left to right). The dislocation movement causes the top half of the crystal to slip by one plane as the former moves from the left (Fig. 4.1a) to the right [Figs. 4.1(b-c)]. The movement of the dislocation across the plane eventually causes the top half of the crystal to move with respect to the bottom half by a *unit slip step* [see Fig. 4.1(d)].

Strengthening Mechanisms By reference to the discussion in the preceding paragraph, we may conclude that a metallic material can be strengthened by obstructing the dislocations movement. This strengthening objective can be achieved by

Z. Huda, *Mechanical Behavior of Materials*, Mechanical Engineering Series,
https://doi.org/10.1007/978-3-030-84927-6_4

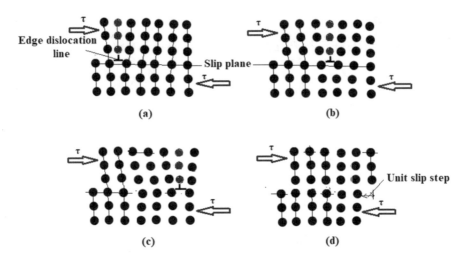

Fig. 4.1 Movement of edge dislocation in a crystal − (**a**) undeformed crystal, (**b,c**) dislocation moved in the crystal, and (**d**) deformed crystal; (τ = shear stress; ⊥ = edge dsilocation line)

introducing obstacles in the motion path of dislocations, such as by introducing interstitial atoms or grain boundaries, to "pin" the dislocations. In the presence of an obstacle, a greater stress is required to move dislocations; which results in strengthening of the metal. For example in cold working, as a material is plastically deformed, more dislocations are produced and intermingle with each other thereby impeding dislocation movement. Thus a higher stress is required to move dislocations thereby increasing the strength. The principal strengthening mechanisms in metals/alloys include: strain hardening, solid-solution strengthening, grain-boundary strengthening, precipitation strengthening, and dispersion strengthening. These strengthening mechanisms are explained in detail in the following sections.

4.2   Grain-Boundary Strengthening

4.2.1   The Evolution of Grained Microstructure

Prior to discussing the grain-boundary strengthening mechanism, it is important to gain a clear understanding of the formation of grained microstructure in a metal. The solidification of a metal is an important industrial process since the processing of most metals and alloys involve melting followed by solidification (*e.g.* casting, welding, etc.). In general, when a metal begins to solidify, multiple crystals begin to grow in the melt and a polycrystalline (more than one crystal) solid forms. The solidification of a polycrystalline material involves the following four stages: (I) *nucleation*: the formation of stable nuclei in the melt (Fig. 4.2a), (II) *crystal growth:* the growth of nuclei into crystals (Fig. 4.2b), (III) the formation of randomly

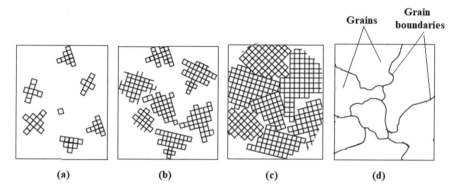

Fig. 4.2 (**a**) Nucleation of crystals, (**b**) crystal growth, (**c**) formation of randomly oriented crystals/grains, and (**d**) grained microstructure as seen under a microscope

oriented crystals (Fig. 4.2c), and (IV) the formation of grained microstructure with grain boundaries (Fig. 4.2d).

Nucleation refers to the appearance of a crystal and the beginning of its growth. The point where nucleation occurs, is called the nucleation point. At the solidification or freezing temperature, atoms of a liquid (e.g. molten metal) begin to bond together at the nucleation points and start to form crystals. The final sizes of the individual crystals depend on the number of nucleation points. The crystals grow by the progressive bonding of atoms until they impinge upon adjacent growing crystals. In a polycrystalline material, each crystal is usually referred to as a grain. It means that a grain is a crystal without smooth faces because its growth was obstructed by contact with another grain or a boundary surface. The interface formed between grains is called a grain boundary. Rapid cooling generally results in more nucleation points and hence larger grain-boundary area resulting in a fine grained microstructure. On the other hand, slow cooling is associated with lesser nucleation points resulting in a coarse-grained microstructure.

4.2.2   Grain-Boundary Strengthening – Hall-Petch Relationship

Grain-boundary strengthening is the strengthening of a polycrystalline material owing to the presence of grain boundaries in its microstructure. A grain boundary may be considered as an array of dislocations (Huda 1991). When a solid is under a shear stress, dislocations tend to move through the lattice (see Fig. 4.1). As a dislocation approaches a grain boundary, the former is unable to easily cross the latter into the adjacent grain. Dislocations are not able to propagate beyond the boundaries of a grain, since adjoining grains will not in general have their slip planes suitably oriented. As a result, the dislocations 'pile up' against the grain boundaries (see Fig. 4.3). In order for the dislocation to enable to cross the grain boundary, a greater stress is needed to be applied thereby increasing the strength of the material.

Fig. 4.3 Dislocations 'pile up' within a grain (d = grain diameter)

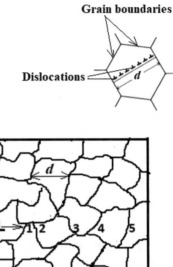

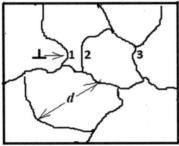 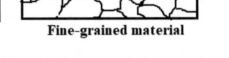

Coarse-grained material   **Fine-grained material**

Fig. 4.4 Dislocations have to cross less (3) grain boundaries in a coarse grained sample, and cross more (5) grain boundaries in a fine-grained material (d = grain diameter)

Fig. 4.4 illustrates the effect of grain size on the strength; here a dislocation has to cross a smaller number of grain boundaries in a coarse-grained metal sample; whereas in the same metal's sample with a fine grain size, the dislocation has to cross a greater number of grain boundaries. It means that by increasing the grain-boundary area, the strength of a metal can be increased. This strengthening mechanism is called *grain-boundary strengthening*; which leads us to establish the following rule of thumb: finer the grain size, stronger the material. However, this rule of thumb holds good for applications of the material at low and moderate temperatures (e.g. Huda 2007).

Hall-Petch Relationship It has been experimentally proved by Petch and Hall that the yield strength of a polycrystalline material varies inversely as the grain size according to the following mathematical relationship (Li, et al. 2016):

$$\sigma_{ys} = \sigma_0 + \frac{K}{\sqrt{d}} \qquad (4.1)$$

where σ_{ys} is the yield strength of the polycrystalline sample, MPa; σ_0 is the yield strength of the single crystal (material with no grain boundaries), MPa; K is the constant referring to the relative hardening contribution of grain boundaries; and d is the average grain diameter, mm (see Examples 4.1, 4.2, 4.3, and 4.4). Eq. 4.1 is known as Hall-Petch relationship −a linear equation in the slope-intercept form. If

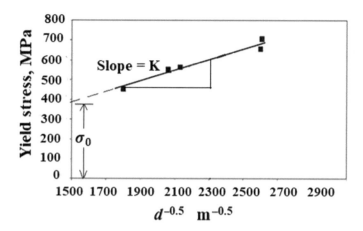

Fig. 4.5 Hall-Petch graphical plot for cryo-milled aluminum

a graph of yield stress σ_{ys} is plotted versus $d^{-\frac{1}{2}}$, a straight line is obtained having a slope = K, and intercept = σ_0 (see Fig. 4.5).

4.3   Strain Hardening

Strain hardening, also called *work hardening*, involves an increase in hardness and strength of a metal due to cold working (see Chap. 2, Sect. 2.5). The strain-hardening strengthening mechanism can be explained in terms of obstructions in the movement of dislocations as follows. As a metallic material is deformed during a cold-working operation, dislocations moving through the material intermingle with one another thereby suddenly increasing the number of dislocation-dislocation interactions. Hence, a greater stress is required to move a dislocation through the intermingled dislocations. This strengthening mechanism is called *strain hardening*. Figure 4.6 illustrates the intermingling of high-density dislocations and elongation of grains resulting from a cold rolling operation.

Strain hardening of a material results in a significant increase in its dislocation density. The *dislocation density* is defined as the number of dislocations in a unit area of the material's sample. The dislocation density can be mathematically expressed by (Callister 2007):

$$\rho_D = \frac{n}{A} \tag{4.2}$$

where ρ_D is the dislocation density, # /cm^2; n is the number of dislocations (or etch pits); and A is the area of the etched sample of the material, cm^2 (see Example 4.5).

Fig. 4.6 Schematic of
intermingling of
dislocation due to cold
rolling (strain hardening)

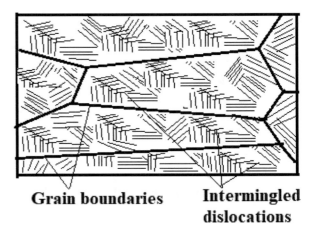

Grain boundaries   **Intermingled
                                  dislocations**

Strain hardening of a metallic material results in a sudden rise in the dislocation density; which in turn results in an increase in hardness and the flow stress of the material. For many metals, the flow stress may be mathematically related to the dislocation density by (Schaffer et al. 1999):

$$\sigma_{flow} = \sigma_0 + k\sqrt{\rho_D} \qquad (4.3)$$

where σ_{flow} is the flow stress, MPa; ρ_D is the dislocation density, dislocations mm^{-2}; and σ_0 and k are the constants for the material (see Examples 4.6 and 4.7).

The strength of a cold-worked metal strongly depends on the degree of deformation (cold working); which is generally expressed as percent cold work (% CW), as follows:

$$\%CW = \left(\frac{A_0 - A_{Cw}}{A_0}\right) \times 100 \qquad (4.4)$$

where A_0 is the original cross-sectional area, and A_{CW} is the cross-sectional area of the cold-worked material. The variations in the tensile mechanical properties with the percent cold rolling of AISI 304 N austenitic stainless steel (ASS) is illustrated in Fig. 4.7 (see Example 4.8). It is evident in Fig. 4.7 that the yield strength and the tensile strength of AISI 304 N ASS increases with % CW, but the ductility (% elongation) decreases with percent cold work (% CW).

Many work-hardened (cold worked) metals and alloys have stress-strain curves which can be represented by a mathematical relationship involving strain-hardening exponent (*n*) and strength coefficient (*K*) (see Chap. 7).

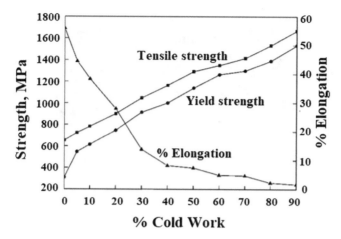

Fig. 4.7 Dependence of tensile properties on % CW of AISI 304 N austenitic stainless steel

4.4   Solid-Solution Strengthening

The presence of solute atoms in a crystal lattice causes some distortion in the lattice; hence greater shear stress is required to move dislocations through the lattice. The increase in the strength of a solid-solution alloy due to the presence of solute, is called *solid solution strengthening*. It occurs when the atoms of the solute form a solid solution with the solvent (base metal), but there is still only one phase; this is why an alloy is generally stronger than pure metal. For example, the tensile strength of commercially pure aluminum is around 30 MPa, but the tensile strength of an aluminum alloy containing 2% Mg is around 75 MPa. In general, higher the concentration of an alloying element, greater will be its strengthening contribution to the alloy.

There are two types of solid solution: (1) substitutional solid solution and (2) interstitial solid solution. In the formation of a substitutional solid solution, if atom(s) of solvent element were to be substituted by solute atom(s); the host lattice would experience a misfit stress associated with the difference in the atomic radii of the solute and solvent (see Fig. 4.8a). An example of substitutional solid solution is iron-cobalt (Fe-Co) alloy; here atomic radii difference in the atomic radii of Fe and Co is small. An interstitial solid solution is formed when small-sized solute atoms are dissolved into interstices of the solvent lattice thereby creating a stress field (Fig. 4.8b). For example small-sized carbon atoms can diffuse into iron lattice to form interstitial solid solution. The strengthening caused by dissolving carbon in BCC iron is more pronounced as compared to FCC iron; this is because the stress field surrounding the solute (carbon) atom in BCC iron is non-symmetrical. On the other hand, the strengthening potential for carbon in FCC iron is less since the strain field surrounding the interstitial (solute) atom site is symmetrical.

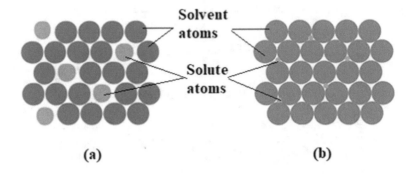

Fig. 4.8 Solid solution strengthening; (a) substitutional solid solution, (b) interstitial solid solution

The strengthening contribution can be measured as the lattice misfit strain ($\varepsilon_{lattice}$), expressed as (Mohri and Suzuki 1999):

$$\varepsilon_{lattice} = k\frac{da}{dc} \tag{4.5}$$

where k, the constant of proportionality, is the reciprocal of lattice parameter ($k = \dfrac{1}{a}$), a is the lattice parameter, and c is the solute concentration in the solid solution.

4.5   Precipitation Strengthening

Precipitation strengthening, also called *coherent precipitate strengthening* or *age hardening*, refers to the increase in the strength of an alloy due to the presence of coherent precipitates (uniformly dispersed particles) within the microstructure. Precipitation strengthening heat-treatment produces coherent precipitates that hinder dislocation motion thereby strengthening the alloy. A coherent precipitate is formed when solute atoms concentrate in certain regions in a crystal lattice, but the solute atoms are still occupying sites in the matrix (parent metal) lattice, and there is no distinct boundary between the main lattice and the *precipitate* (see Fig. 4.9).

Examples of precipitation-strengthened alloys include: maraging steels, age-hardenable aluminum alloys, nickel-base superalloys, and the like (Huda 2007, 2009) (see also Ch. 5). For instance, precipitation strengthening occurs by the formation of coherent θ' precipitates in an aluminum-copper (*Al*–4 wt% Cu) alloy according to the following phase-transformation reaction:

$$\underset{[\text{in } Al \text{ lattice}]}{Cu} + \underset{[Al \text{ lattice}]}{2Al} \rightarrow \underset{[\text{coherent precipitate}]}{\theta'} \tag{4.6}$$

where the coherent precipittae θ' is the intermetallic compound: $CuAl_2$ (see Fig. 5.7 in Chap. 5). This author has conducted heat treatment experiments and has shown a

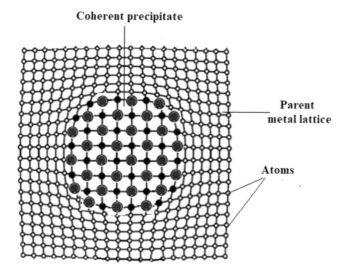

Fig. 4.9 Precipitation strengthening showing coherent precipitate

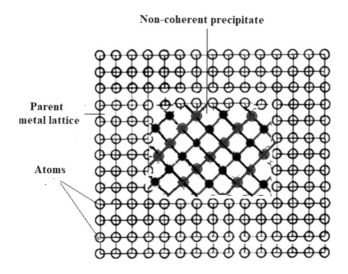

Fig. 4.10 Dispersion strengthening showing non-coherent precipitate

good θ' precipitation in the 2017 aluminum alloy (Huda 2009). Figure 4.10(a) shows that there is a lattice mismatch due to precipitation strengthening. In superalloys, the % lattice misfit is computed by (Stoloff 1987):

$$\%\text{Lattice mismatch} = \frac{a_{\gamma'-a_\gamma}}{a_\gamma} \times 100 \qquad (4.7)$$

where a_γ is the lattice parameter of the nickel matrix (FCC, γ) phase, nm; and $a_{\gamma'}$ is the lattice parameter of the second-phase gamma-prime (γ') phase, nm (see Examples 4.9 and 4.10).

4.6 Dispersion Strengthening – *Mechanical Alloying*

Dispersion strengthening, refers to the strengthening of an alloy due to the presence of dislocation pinning sites that are achieved by nano-sized widely-dispersed non-coherent particles in the microstructure (see Fig. 4.10). There must be at least two phases in the microstructure of a dispersion-strengthened alloy. The continuous phase, is called the matrix whereas the second phase is the dispersed (non-coherent) precipitate. A non-coherent precipitate has no relationship with the crystal structure of the matrix (Fig. 4.10).

In order to produce a dispersion-strengthened alloy, the matrix (parent metal) should be ductile and soft, whereas the precipitate (second phase) should be strong. The dispersed (non-coherent) precipitate hinders dislocation motion, whilst the matrix provides some ductility to the overall alloy. Notable examples of dispersion-strengthened alloys include: mechanically-alloyed dispersion-strengthened superalloys, oxide-dispassion-strengthened (ODS) alloys [e.g. thoria-dispersed (TD) nickel, sintered alumina powder (SAP), etc.], Co-WC composite, and the like.

4.7 Calculations – *Worked Examples (Solved Problems)*

Example 4.1 Determining the Average Grain Diameter using Hall-Petch Relation's Plot By reference to Fig. 4.5, determine the average grain diameter of cryo-milled aluminum that corresponds to 600 MPa yield strength.

Solution From Fig. 4.5, the 600 MPa yield strength corresponds to $d^{-0.5} = 2300\,\text{m}^{-0.5}$

$$d^{-\frac{1}{2}} = 2300\,\text{m}^{-\frac{1}{2}}$$

$$\frac{1}{\sqrt{d}} = 2300\,\frac{1}{\sqrt{m}}$$

$$\frac{1}{d} = (2300)^2\,\frac{1}{m}$$

$$d = \frac{1}{2300^2}\,m = \frac{1}{5290000}\,m = 189 \times 10^{-9}\,\text{m} = 189\,\text{nm}$$

The average grain diameter corresponding to the 600 MPa yield strength = 189 nanometer.

Example 4.2 Determination of the constant in Hall-Petch relation by graphical plot By reference to Fig. 4.5, determine the constants K and σ_0 in the Hall-Petch relationship for cryo-milled aluminum.

Solution In order to determine the values of the slope K and the intercept σ_0, the graphical plot (Fig. 4.5) is reproduced as shown in Fig. 4.11.

By reference to Fig. 4.11,

$$\text{Slope} = K = \frac{\Delta \sigma_{ys}}{\Delta d^{-\frac{1}{2}}} = \frac{600 - 470}{2300 - 1810} = \frac{130\,MPa}{490\ m^{-\frac{1}{2}}} = \frac{130\,MPa}{490\left(10^3\,mm\right)^{-\frac{1}{2}}} = 0.26 \times 31.62 = 8.2\,MPa\ \ \sqrt{mm}$$

$$\text{Intercept} = \sigma_0 = 385\,MPa$$

The Hall-Petch relation's constants for cryo-milled *Al* are: $K = 8.2\,MPa\ \sqrt{mm}$, $\sigma_0 = 385$ MPa

Example 4.3 Calculating σ_0 and K in the Hall-Petch Equation A steel sample with an average grain size of 0.011 mm has a yield strength of 375 MPa. The same steel after grain-growth annealing produced another sample with an average grain size of 0.016 mm having a yield strength of 341.3 MPa. Calculate the Hall-Petch relation's constants: K and σ_0.

Solution Sample 1: d_1 = 0.011 mm, σ_{ys1} = 375 MPa; Sample 2: d_2 = 0.016 mm, σ_{ys2} = 341.3 MPa

By using Equation 4.1 for sample 1,

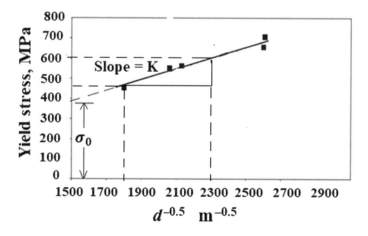

Fig. 4.11 Determination of K and σ_0 in the Hall-Petch relation by graphical plot

$$\left(\sigma_{ys}\right)_1 = \sigma_0 + \frac{K}{\sqrt{d_1}}$$

$$375 = \sigma_0 + \frac{K}{\sqrt{0.011}} \tag{E-4.3a}$$

By using Equation 4.1 for sample 2,

$$341.3 = \sigma_0 + \frac{K}{\sqrt{0.016}} \tag{E-4.3b}$$

By performing the subtraction of equations: (E-4.3a – E-4.3b),

$$375 - 341.3 = \frac{K}{\sqrt{0.011}} - \frac{K}{\sqrt{0.016}}$$

$$K\left(\frac{1}{\sqrt{0.011}} - \frac{1}{\sqrt{0.016}}\right) = 33.7$$

$$K = 16\,MPa\ \sqrt{mm}$$

By substituting the value of K in Equation E-4.3b,

$$341.3 = \sigma_0 + \frac{16}{\sqrt{0.016}}$$

$$\sigma_0 = 341.3 - \frac{16}{0.126} = 341.3 - 127 = 214\,MPa$$

The constants are: $\sigma_0 = 214$ MPa, $K = 16\,MPa\ \sqrt{mm}$

Example 4.4 Calculating the Yield Strength when the Grain Diameter is known By using the data in Example 4.3, calculate the yield strength of the steel sample with an average grain diameter of 0.013 mm.

Solution $K = 16\,MPa\ \sqrt{mm}$, $\sigma_0 = 214$ MPa, $d = 0.013$ mm, $\sigma_{ys} = ?$

By using Equation 4.1

$$\text{Yield strength} = \sigma_{ys} = \sigma_0 + \frac{K}{\sqrt{d}} = 214 + \frac{16}{\sqrt{0.013}} = 214 + \frac{16}{0.114} = 354.3\,MPa$$

Example 4.5 Calculating the Dislocation Density by Etch-Pit Counts An etched sample of a deformed metal contains 8×10^6 dislocations (etch pits) in a 4 mm^2 area. Compute the dislocation density in the metal's sample.

Solution $n = 8 \times 10^6$ dislocations, $A = 4$ mm^2 $= 4 \times (10^{-1}$cm$)^2 = 4 \times 10^{-2}$cm^2, $\rho_D = ?$

By using Equation 4.2,

$$\text{The dislocation density} = \rho_D = \frac{n}{A} = \frac{8 \times 10^6}{4 \times 10^{-2}} = 2 \times 10^8 \, cm^{-2}$$

Example 4.6 Calculating k and σ_0 for the flow stress-dislocation density relationship A cold working operation was performed for a metallic material; as a result the flow stress increased from 3 to 40 MPa with accompanying increase in dislocation density from 10^7 to 10^{10} cm^{-2}. Calculate k and σ_0 for the flow stress-dislocation density relationship.

Solution $(\sigma_{flow})_1 = 3$ MPa, $(\sigma_D)_1 = 10^7$ cm^{-2}, $(\sigma_{flow})_2 = 40$ MPa, $(\sigma_D)_2 = 10^{10}$ cm^{-2}, $\sigma_0 = ?$, k $= ?$,

By using Equation 4.3 for pre-cold working,

$$\left(\sigma_{flow} \right)_1 = \sigma_0 + k \sqrt{\left(\sigma_D \right)_1}$$

$$3\,MPa = \sigma_0 + k\sqrt{10^7 \, cm^{-2}} \tag{E-4.6a}$$

By using Equation 4.3 for post-cold working,

$$\left(\sigma_{flow} \right)_2 = \sigma_0 + k \sqrt{\left(\sigma_D \right)_2}$$

$$40\,MPa = \sigma_0 + k\sqrt{10^{10} \, cm^{-2}} \tag{E-4.6b}$$

By performing the subtraction of equations (E-4.6b) – (E-4.6a), we obtain:

$$37\,MPa = k\left(\sqrt{10^{10} \, cm^{-2}} - \sqrt{10^7 \, cm^{-2}} \right)$$

$$k = \frac{37\,MPa}{\sqrt{10^{10} \, cm^{-2}} - \sqrt{10^7 \, cm^{-2}}} = \frac{37\,MPa}{10^5 \, cm^{-1} - 3162\,cm^{-1}} = 382 \times 10^{-6} \, MPa \quad cm$$

$$k = 3.82 \times 10^{-4} \, MPa \quad cm$$

By substituting the value of k in Equation E-4.6a,

$$3\,MPa = \sigma_0 + k\sqrt{10^7 \, cm^{-2}} = \sigma_0 + 3.82 \times 10^{-4} \, MPa \quad cm \times \sqrt{10^7 \, cm^{-2}}$$

$$3\,MPa = \sigma_0 + 12080 \times 10^{-4} \, MPa = \sigma_0 + 1.2\,MPa$$

$$\sigma_0 = 3 - 1.2 = 1.8\,MPa$$

The constants are: $k = 3.82 \times 10^{-4}$ MPa - cm, and $\sigma_0 = 1.8$ MPa

Example 4.7 Calculating the Flow Stress when the Dislocation Density is Known By using the data in Example 4.6, calculate the flow stress of the cold-worked metal after further deformation that results in increase of dislocation density to 10^{11} cm^{-2}.

Solution $k = 3.82 \times 10^{-4}$ MPa - cm, and $\sigma_0 = 1.8$ MPa, $\rho_D = 10^{11}$cm^{-2}, $\sigma_{flow} = ?$

By using Equation 4.3,

$$\sigma_{flow} = \sigma_0 + k\sqrt{\rho_D} = 1.8 + \left(3.82 \times 10^{-4} \times \sqrt{10^{11}}\right) = 1.8 + 120.8 = 122.6 \, MPa$$

The flow stress corresponding to dislocation density to 10^{11} cm^{-2} = 122.6 MPa

Example 4.8 Finding the Tensile Properties by Calculating % CW and by using Plot A 3-mm-thick sheet made of AISI 304N ASS was cold rolled to 1-mm thickness with no change in width. Determine the tensile mechanical properties of the cold-worked stainless steel.

Solution $A_0 = t_0 \cdot w = 3 \times w \, mm^2 = 3w$

$$A_{CW} = t_f \cdot w = 1 \times w \, mm^2 = w$$

By using Equation 4.4,

$$\%CW = \left(\frac{A_0 - A_{Cw}}{A_0}\right) \times 100 = \left(\frac{3w - w}{3w}\right) \times 100 = 66.7$$

By reference to Fig. 4.7, the 67 % cold work corresponds to the following mechanical properties: Yield strength = 1,200 MPa, Tensile strength = 1,380 MPa, Elongation = 5 %

Example 4.9 Calculating the lattice parameter of the γ′phase in a Strengthened Superalloy A superalloy's microstructure comprises of nickel matrix (FCC, γ) phase and the second-phase gamma-prime (γ′) phase. The γ phase has a lattice parameter of 0.358 nm. Calculate the lattice parameter of the γ′phase, if the percent lattice mismatch of the alloy is –0.2.

Solution $a_\gamma = 0.358$ nm, % Lattice mismatch = –0.2, $a_{\gamma'} = ?$

By using Equation 4.7,

$$\% \text{Lattice mismatch} = \frac{a_{\gamma'-a_\gamma}}{a_\gamma} \times 100$$

$$-0.2 = \frac{a_{\gamma'-0.358}}{0.358} \times 100$$

$$a_{\gamma'} - 0.358 = -0.000716$$

The lattice parameter of the γ' phase $= a_{\gamma'} = 0.3573\,\text{nm}$

Example 4.10 Calculating the Percent Lattice Mismatch in a Strengthened *Al-Cu Alloy* The lattice parameter of aluminum (FCC lattice) is 0.404 nm. A precipitation-strengthened aluminum-copper alloy's microstructure comprises of FCC γ phase and the precipitated θ' phase. The lattice parameter of θ' phase is 0.29 nm. Calculate the percent lattice mismatch for γ/θ'.

Solution $a_\gamma = 0.404$ nm, $a_{\theta'} = 0.290\,\text{nm}$, % Lattice mismatch = ?

By using the modified Equation 4.7 for the aluminum-copper alloy, we obtain:

$$\% \text{Lattice mismatch} = \frac{a_{\theta'-a_\gamma}}{a_\gamma} \times 100 = \frac{0.290 - 0.404}{0.404} \times 100 = -28.2$$

Questions and Problems

4.1. (MCQs). Underline the most appropriate answers for each of the following questions:

(a) Which type of strengthening requires a higher stress to cross an array of dislocations?
   (i) strain hardening, (ii) grain-boundary strengthening, (iii) solid solution strengthening.

(b) Which type of strengthening requires a greater stress to move a dislocation through the large number of intermingled dislocations?
   (i) strain hardening, (ii) grain-boundary strengthening, (iii) precipitation strengthening.

(c) Which type of strengthening is achieved by the presence of coherent precipitates?
   (i) dispersion strengthening, (ii) solid-solution hardening, (iii) precipitation strengthening.

(d) Which type of strengthening is achieved by the presence of non-coherent precipitates?

(i) dispersion strengthening, (ii) strain hardening, (iii) precipitation strengthening.

    (e) Which type of strengthening results to stress field caused by the presence of solute atoms?

       (i) strain hardening, (ii) solid-solution hardening, (iii) precipitation strengthening

4.2. Explain the role of dislocation movements in plastic deformation of a solid with the aid of sketches.

4.3. (a) What is the basic principle of strengthening mechanism?

    (b) List the various strengthening mechanisms, and explain any of them.

4.4. (a) Why is an alloy generally stronger than pure metal?

    (b) Draw diagrams showing interstitial solid solution and substitutional solid solutions.

    (c) Why is strengthening by dissolving carbon in BCC iron is more pronounced as compared to dissolving carbon in FCC iron?

4.5. Explain grain-boundary strengthening with the aid of diagram(s).

4.6. Explain strain hardening with reference to dislocation density.

4.7. Draw diagrams illustrating coherent precipitates and non-coherent precipitate.

4.8. Differentiate between precipitation strengthening and dispersion strengthening, giving examples for each type of strengthening.

4.9. The yield strength of a sample of annealed aluminum with an average grain size of 0.037 mm is 36 MPa. The yield strength of a single-crystal sample of the metal is $\sigma_0 = 25$ MPa. Calculate the yield strength of a sample of aluminum with average grain size of 0.05 mm.

4.10. Refer to Fig. 4.4. What is the average grain diameter (in nanometer) of cryo-milled aluminum that corresponds to the highest yield strength in the graphical plot?

4.11. By using the data in Problem 4.10, draw a graphical plot for $d^{-\frac{1}{2}}$ versus σ_{ys}; and hence graphically determine the constants K and σ_0 in the Hall-Petch relationship for the metal.

4.12. An etched sample of a strain-hardened metal contains 10^7 etch pits (dislocations) in a 3 mm² area. What is the dislocation density in the metal sample?

4.13. A precipitation-strengthened nickel-base alloy's microstructure comprises of FCC γ phase (matrix) and the precipitated γ' phase having a lattice parameter of 0.3568 nm. Calculate the percent lattice mismatch. Nickel (FCC lattice) has the atomic radius of 0.1246 nm

4.14. By using the data in Problem 4.11, calculate the flow stress of the metal. The constants for the flow stress-dislocation density relation are: $k = 1.7 \times 10^{-3}$ MPa-cm and $\sigma_0 = 0.05$ MPa.

4.15. A 5-mm-thick sheet made of made of AISI 304 N ASS is cold rolled to 3.5-mm thickness with no change in width. Determine the tensile mechanical properties of the steel.

References

Argon AS (2007) Strengthening mechanisms in crystal plasticity. Oxford University Press, Oxford

Callister WD (2007) Materials science and engineering: an introduction (7th edition). John Wiley & Sons, Inc., New York

Huda Z (1991) PhD Thesis, Materials Tech. Department. Brunel University, London

Huda Z (2007) Development of heat treatment process for a P/M superalloy for turbine blades. Mater Des 28(5):1664–1667

Huda Z (2009) Precipitation strengthening and age hardening in 2017 aluminum alloy for aerospace application. Eur J Sci Res 26(4):558–564

Li Y, Bushby AJ, Dunstan DJ (2016) The Hall–Petch effect as a manifestation of the general size effect. Proceedings of the Royal Society, London

Mohri T, Suzuki T (1999) Solid solution hardening by impurities. In: Briant CL (ed) Impurities in engineering materials. CRC Press, Boca Raton, pp 259–270

Schaffer JP, Saxena A, Antolovich SD, Sanders TH Jr, Warner SB (1999) The science and design of engineering materials. WCB-McGraw Hill Inc., New York

Stoloff NS (1987) Fundamentals of strengthening. In: Sims CT, Stoloff NS, Hagel WC (eds) Superalloys-II. John Wiley & Sons Inc, New York

Chapter 5
Materials in Engineering

5.1 Materials and Engineers

Today's modern world is heavily dependent on materials in engineering. Engineers are tasked to select materials on the basis of their mechanical behavior and other properties for designing machines and structures. Thus, machine elements are designed by selecting suitable materials with the right mechanical behavior and performance to meet the needs of modern technology. It is therefore very important for an engineer to possess a good knowledge of materials in engineering. All engineering materials may be classified into four basic categories: (1) metals/alloys, (2) polymers/plastics, (3) glasses and ceramics, and (4) composites. Additionally, semiconductors and advanced materials have revolutionized the world in terms of automation and speed.

5.2 Classification of Materials in Engineering

Engineers are expected to select materials in designing machines and/or their elements. There are hundreds of materials available to choose by a machine designer. Most engineering materials fall into one of the following four classes of materials: metals/alloys, ceramics/glasses, polymers/plastics, and composites. Additionally, there are a number of classes of materials based on their applications; these classes include: electronic materials (semiconductors), smart materials, biomaterials, nanomaterials, and the like. Materials in engineering may be classified into four classes: (a) metals, (b) ceramics, (c) polymers, (d) composites (see Fig. 5.1). Each of these classes of materials is explained in the following sections.

© The Author(s), under exclusive license to Springer Nature Switzerland AG 2022
Z. Huda, *Mechanical Behavior of Materials*, Mechanical Engineering Series,
https://doi.org/10.1007/978-3-030-84927-6_5

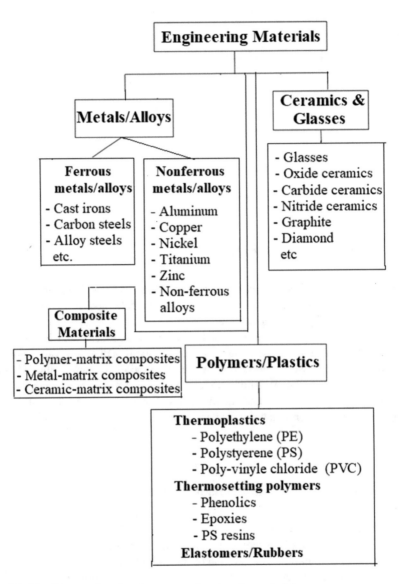

Fig. 5.1 Classification chart showing groups of engineering materials

5.3   Metals and Alloys

Metals and alloys are inorganic materials having a crystalline structure as well as strength and ductility. *Metals* are electropositive (chemical) elements with high electrical and thermal conductivities. Examples of metals include iron, aluminum, zinc, copper, and the like. *Alloys* are composed of one or more metals; they may also contain some non-metallic elements. Common examples of alloys include: steels, brasses, bronzes, superalloys, and the like. Steels are iron-based alloys containing (up to 1.8 wt%) carbon and small amounts of impurities; alloy steels contain substantial amounts of alloying elements (*e.g.* nickel, chromium, vanadium, etc). Brasses are alloys of copper and zinc. Metals and alloys find wide applications in structural engineering, energy sector, aerospace, biomedical, electrical, transport, beverage industries, and the many more sectors.

Metals and alloys may be divided into two categories: *ferrous alloys* and *nonferrous alloys* . *Ferrous* alloys contain a large percentage of iron ; examples of ferrous alloys include: cast irons, carbon steels, alloy steels, and the like. *Non-ferrous alloys* do not contain iron or contain only small amount of iron; examples of nonferrous alloys include: aluminum and their alloys, copper and their alloys, nickel and their alloys, etc. (see Fig. 1.1).

FERROUS ALLOYS

5.4   Cast Irons

5.4.1   *Characteristics and Applications of Cast Irons*

Cast irons (*CI*) are ferrous alloys that contain more than 2% carbon and 0.5% or more silicon. They are the least expensive of all engineering metallic materials. The characteristics of cast irons include: low equipment and production costs, ready availability, good machinability without burring, easy to cast into complex shapes, excellent wear resistance, high hardness, and high damping ability. Cast irons find wide applications as automotive components (*e.g.* brake disc, clutch plates, cylinder blocks and heads, piston, liner, etc) and other engineering components (*e.g.* gears, pipes, water pipes, flywheels, man-hole cover, etc.). The carbon equivalent (*CE*) of a cast iron enables us to distinguish between the two principal types of cast iron: *gray CI* and *white CI*. The carbon equivalent (CE) is numerically defined as:

$$CE\left(\text{in wt.%}\right) = C + \frac{Si + P}{3} \tag{5.1}$$

where *C* is the wt. % carbon, *Si* is the wt. % silicon, and *P* is the wt. % phosphorous. For optimum metallurgical properties in gray CI, the value of CE should not be less than 4.3. The values of CE lower than 4.3 are desired in white *CI* (see *Example 5.1*).

5.4.2   *Types of Cast Irons*

Cast irons may be divided into seven types: (a) gray CI, (b) white CI, (c) ductile or spheroidal graphitic iron, (d) malleable iron, (e) alloyed CI, (f) compacted graphite CI, and (g) austempered ductile iron (ADI). These types of cast iron are briefly described in the following paragraphs.

Gray Cast Iron   Gray CI is so named because its fractured surface appears gray due to the presence of graphite. They are characterized as randomly-oriented graphite flakes in their microstructures (see Fig. 5.2a). *Gray CI* possess good wear resistance and machinability as well as excellent damping capacity. This is why, gray CI finds applications in internal combustion engines' components and the base structures of heavy machines that are exposed to vibration.

White Cast Iron   White cast iron is so named because its fractured surface appears white due to the absence of graphite *i.e.* carbon is in the form of iron carbide – cementite (*Fe₃C*). White CI is a hard and brittle material with poor machinability; however its compressive strength and wear resistance are good. White *CI* finds applications in pumps, grinding tools, man-hole covers, etc.

Ductile or Nodular Cast Iron   Ductile cast iron, also called nodular cast iron or spheroidal graphitic (SG) iron, is so named due to the presence of graphite in the form of nodules/spheroids in its microstructure (see Fig. 5.2b). A typical composition of SG iron is as follows: Fe–3.2 wt.% C–2.2 wt.% Si–0.3 wt.% Mn–0.04 wt.% Mg–0.01 wt.% P–0.01 wt.% S. Nodular CI possess a good combination of moderately high tensile strength and ductility; these mechanical behaviors renders nodular *CI* suitable for applications in pipes for water and sewerage lines as well as in automotive components (*e.g.* cylinder heads, brake-drums and discs). In order to ensure a good mechanical behavior, there must be a large number of round-shaped nodules in the microstructure.

The graphite nodules count per unit volume (N_V) can be computed by (Huda, 2020):

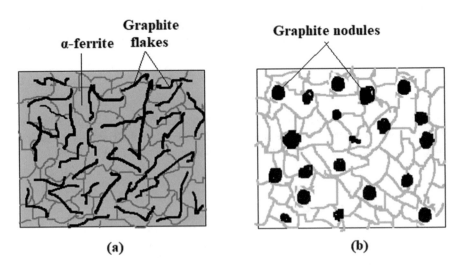

Fig. 5.2   Schematic microstructures of gray cast iron (**a**), and ductile cast iron (**b**)

$$N_V = \sqrt{\frac{\pi}{6 f_g}} \left(\alpha N_A\right)^{1.5} \qquad (5.2)$$

where N_V is the nodules count per unit volume, nod/mm³; f_g is the fraction of graphite; α is a parameter dependent on the width of the size distribution of the nodules ($\alpha = 1$–1.25); and N_A is the nodules count per unit area, nod/mm² (see *Example 5.2*).

Compacted Graphite (CG) Iron   CG iron is characterized by randomly-oriented graphite flakes that have rounded edges and shorter & thicker as compared to those in gray CI. This microstructural feature of *CG iron* imparts a good fracture toughness by suppressing crack initiation and propagation. *CG iron* also has superior mechanical properties relative to gray CI and better thermal conductivity relative to *SG* iron (Jones, 2020).

Malleable Cast Iron   *Malleable cast iron* is basically a white *CI* that has been heat treated; which results in the decomposition of cementite to free nodules of graphite. Malleable CI has improved ductility combined with a fairly good tensile strength. There are three types of malleable CI: (a) white-heart CI, (b) black-heart CI, and (c) pearlitic CI. Malleable CIs are used in thin-section castings and the components with good formability, impact toughness, and machinability.

Alloyed Cast Iron   *Alloyed cast irons* are the cast irons alloyed with nickel, chromium, and molybdenum; its microstructure contains carbides with graphite flakes in a matrix of either pearlite or bainite. *Alloyed CIs* having high hardness (where the matrix is bainitic instead of pearlitic) and are widely used in manufacturing finishing stands of rolling mills.

Austempered Ductile Irons (ADI)   *ADI* are the cast irons that have been heat treated by austenizing at 950 °C followed by austempering at 350 °C for 1 h to obtain bainitic structure. A typical composition of ADI is as follows: Fe–3.52C–2.5 1Si–0.49Mn–0.15Mo–0.31Cu. *ADIs* possess good fatigue strength and damping capacity best suitable for automotive applications.

5.4.3   *Mechanical Properties of Cast Irons*

A general description of the mechanical properties has been given in the previous paragraph.

Some important mechanical properties of gray CI, CG iron, and SG irons are shown in Table 5.1.

Table 5.1 Some mechanical properties of cast irons

Properties	Tensile strength (MPa)	Young's modulus (GPa)	% Elongation	Fatigue strength (MPa)	Hardness (HB)
Gray CI	250–270	105	0	110	200
CG iron	450–500	145	1.5	200	220–240
Ductile CI	750	160	2.5	250	220–250

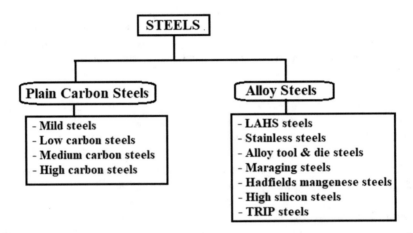

Fig. 5.3 Classification of steels (LAHS low alloy high strength, TRIP TRansformation Induced Plasticity)

5.5 Steels

5.5.1 Steels' Definition, Classification, and Designation Systems

Steels are ferrous alloys containing less than 2 wt% carbon; there may be small amounts of impurities and/or alloying elements in steels. Steel is produced by refining pig iron by oxidation of impurities in a steel-making furnace. Steels are broadly classified into two groups: (1) carbon steels, and (2) alloy steels (see Fig. 5.3). Steels are generally designated based on SAE (Society of Automotive Engineers) and AISI (American Iron & Steel Institute) numbering systems. The steel designation systems generally use four-digit numbers; the first two digits indicate the major alloying elements, and the last two digits refer to the wt% carbon in the steel. The digit series 1x × × indicates carbon steels. The 10xx series refers to carbon steels containing 1.00% manganese (max.); the 11xx series represents resulfurized carbon steels; 12xx series refers to resulfurized and rephosphorized carbon steels; and the 15xx series indicates non-resulfurized high-manganese (up-to 1.65 wt.% Mn) carbon steels possessing good machinability. For example, an AISI 1030 steel is a plain-carbon steel with 0.30 wt.% carbon. The AISI-SAE system classifies alloy steels using first digit for a specific alloying element (see Sect. 5.5.3.1).

5.5.2 Carbon Steels

5.5.2.1 Classification and Applications of Carbon Steels

Carbon steels are the steels that contain 0.05% to 1.5% carbon; and which generally contain less than 0.5% wt% silicon and 1.5% wt% manganese. Carbon steels may be divided into four types: (1) mild steels (MS), (2) low-carbon steels, (3) medium carbon steels, and (4) high carbon steels. These types of steels are described in the following paragraphs.

A *mild steel*, is a very low-carbon steel that contains not more than 0.05 wt% carbon. Mild steels are ductile and have properties similar to wrought iron. They cannot be modified by heat treatment; they find applications in non-critical components and structures.

Low carbon steels contain 0.05 to 0.2 wt% carbon. Examples of *low-carbon steels* include: AISI 1004 (0.04% C steel), AISI 1008, AISI 1018 steels, and the like. These steels cannot be effectively heat treated; they have low hardness and strength. *Low-carbon steels* find applications requiring high ductility, such as screws, nails, wires, and the like. *Medium carbon steels* contain 0.2 to 0.8 wt% carbon. Examples of *medium-carbon steels* include: AISI 1025, AISI 1050, AISI 1070, and the like. As compared to low-carbon steels, they have higher hardness and strength. They can be effectively heat treated, are used for manufacturing gears, pylons, pipelines, and the like. *High carbon steels* contain 0.8 to 1.5% wt% carbon. They can be effectively hardened by heat treatment. They are used in making knives, files, chisel, and other cutting tools.

5.5.2.2 Microstructures of Carbon Steels

The microstructures of typical samples of low-carbon steel, medium-carbon steel, and high-carbon steel are shown in Figs. 5.4(a–c). The microstructural phases and microconstituents of carbon steels can be deduced from iron-carbon phase diagram, as follows. When a low-carbon steel is fully annealed (heated above 727 °C to form austenite and then slowly cooled to ambient temperature), the resulting microstructure will comprise of (ferrite + pearlite) (Fig. 5.4a). Similarly, the microstructure of steel containing 0.77% C (a medium-carbon steel) shows pearlite (Fig. 5.4b). The microstructure of high-carbon steel shows (pearlite + cementite) (Fig. 5.4c).

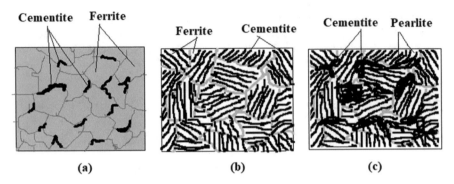

Fig. 5.4 The microstructures of typical samples of low-carbon steel (**a**), medium carbon steel showing pearlite (**b**), and high-carbon steel (**c**)

5.5.2.3   Mechanical Properties of Carbon Steels

The mechanical properties of carbon steels depend on the microstructure; which in turn depend on the composition and processing, particularly heat treatment. Carbon steels contain carbon (C) and other impurities; which include silicon (Si), manganese (Mn), sulfur (S), and phosphorous (P). In addition to C, Si, Mn, S, and P, there may be small amounts of oxygen, nitrogen, and hydrogen in steels. In particular, the presence of significant hydrogen in extremely detrimental to mechanical behavior of steels. The scope of this book does not permit this author to discuss the effects of all impurities in carbon steels; the reader is, therefore advised to refer to literature (*e.g.* Huda, 2020). However, the effects of carbon on steel is explained in the following paragraph.

An increase in the carbon content from 0.01 to 1.5% in carbon steel generally results in an increase in its hardness and strength; but, there is a corresponding decrease in the ductility and toughness with additions of carbon in steels. The effects of carbon content on the strength and ductility of ferrite-pearlite steel is illustrated in Fig. 5.5. It is evident in Fig. 5.5 that the tensile strength (σ_{ut}) of steel varies linearly as its carbon content increases from 0.1 to 0.5 wt.%; beyond which the rise in σ_{ut} follows a curve. However, ductility (% reduction in area) significantly deceases as carbon content increases from 0.1 to 0.7% (see Examples 5.3, 5.4, and 5.5).

5.5.3   *Alloy Steels*

5.5.3.1   Introduction to Alloy Steels

An *alloy steel* is the steel to which one or more alloying elements (up to 50 wt.%) besides carbon have been intentionally added for obtaining the improved mechanical properties. The main alloying elements added to produce an alloy steel include: silicon (Si), manganese (Mn), nickel (Ni), chromium (Cr), molybdenum (Mo),

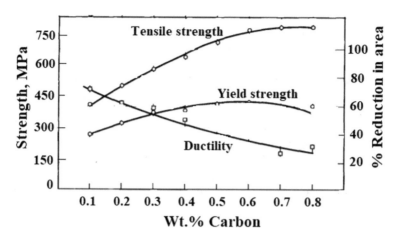

Fig. 5.5 The effects of carbon contents on the mechanical behavior of annealed steels

tungsten (W), vanadium (V), cobalt (Co), titanium (Ti), boron (B), copper (Cu), aluminum (Al), and niobium (Nb). In particular, alloying with Ni, Nb, or V generally results in an increase in hardness and strength of steel without decreasing its ductility.

Alloy steels are classified based on the *AISI* designation system by using the first digit as follows: 2 for Ni steels, 3 for Ni-Cr steels, 4 for Mo steels, 5 for Cr steels, 6 for Cr-V steels, 7 for W-Cr steels, and 9 for Si-Mn steels. For example, a typical alloy steel with AISI-SAE designation number 4340 has the following composition ranges (wt.%): 0.37–0.43 C, 0.2–0.3 Mo, 0.6–0.8 Mn, 1.6–2.0 Ni, 0.7–0.9 Cr, 0.15–0.3 Si, 0.04 S, 0.035 P, balance: Fe (DeGarmo, et al., 2003). The application of an alloy steel strongly depends on the nature and the amount of alloying elements and the heat treatment given to the steel. For example, the 4140 alloy steel is used in shafts, gears, and forgings; 4340 steel is used in aircraft tubes, bushings, and gears, and the like.

5.5.3.2 Types of Alloy Steels and their Mechanical Behaviors

We have learnt in Sect. 5.5.1 that alloy steels may be divided into the following types: (a) low-alloy high-strength steels, (b) stainless steels, (c) alloy tool and die steels, (d) maraging steels, (e) Hadfields manganese steels, (f) high-silicon electrical steels, and TRIP steels (see Fig. 5.3).

Low alloy high strength (LAHS) Steels are the low-carbon steels strengthened by the addition of 1–4 wt% alloying elements (with up to 1.65 wt% manganese); these steels are sometimes strengthened by special rolling and heat-treatment techniques. A typical *LAHS* steel contains about 0.15 wt.% C, 1.65 wt.% Mn and P and S (each <0.035%), and small additions of Cu, Ni, Cr, Mo, Nb, and V. Air hardening *LAHS* steel contains about 4.25 wt.% Ni and 1.25 wt.% Cr. An additions of 0.3 wt.% Mo renders nickel-chrome-molybdenum steels suitable for applications in axles, shafts, gears, connecting rods, and other automotive applications.

Tool and die steels possess high hardness, toughness, hardenability, and the resistance to wear and corrosion. These steels contain Cr, W, Mo, and V; which are carbide formers and stabilize ferrite and martensite. A typical composition of tool steel is Fe–0.8%C–18%W–4%Cr–1%V, and finds application as high-speed steel in making cutting tools.

Hadfield Manganese Steels are the alloy steels that contain 12% to 14% Mn and 1% C. These steels are austenitic at all temperatures and are non-magnetic. These steels are used in pneumatic drill bits, excavator bucket teeth, rock crusher jaws, ball mill linings and railway points and switches. When Hadfields steel is water quenched from 1050 °C to retain carbon in solution, the soft core has a tensile strength of 849 MPa, a ductility of 40% and a Brinell hardness of 200; but after abrasion the surface hardness increases to 550 BHN.

Maraging steels contain high Ni, Co, and Mo. A typical composition of Maraging steel is: Fe–18%Ni–8%Co–5%Mo–0.4%Ti–0.1%Al and up to 0.05% C. Heat treatment involves solution treatment followed by quenching of the austenite to give a BCC martensitic structure. Aging heat treatments can produces finely dispersed precipitates of complex intermetallics (*e.g.* $TiNi_3$). This precipitation-strengthening results in high tensile strengths of around 2000 MPa. These steels are relatively tough; with good corrosion resistance and good weldability. Maraging steels find applications in aircraft undercarriage components, wing fittings, extrusion-press rams, and the like.

Stainless steels (SS) are the corrosion resistant steels containing more than 12% Cr with some nickel (Ni). Stainless steels can be classified into four types: (a) ferritic SS, (b) martensitic SS, (c) austenitic SS, and (d) duplex SS. *Ferritic stainless steels (FSS)* contain 12% to 25% Cr and up to 0.1% C. A typical composition of FSS is as follows: Fe-18%Cr-8%Ni-0.1%C; which is called *18–8 stainless steel*. FSS are ferritic up to the melting temperature; they cannot be quench hardened to produce martensite. *Martensitic stainless steels (*MSS) contain 12% to 25% Cr and 0.1% to 1.5%C. This steel can be heated into the austenite region and quenched to give a martensitic structure. *MSS* have high strength and wear resistance with corrosion resistance; they are suitable for applications in hydroelectric turbines, knife, cutlery, and the like. *Austenitic stainless steels (ASS)* are non-magnetic SS that contain 16 to 26% Cr and up to 35% Ni. *ASS* has higher ductility than other types of SS, and can be cold-worked to produce deep-drawn shapes used in chemical plant, kitchenware, and architectural work. *Duplex stainless steels (DSS)* have a two-phase (duplex) microstructure consisting of ferrite and austenite (see Fig. 5.6a). *DSS* contain Cr, Mo, and Ni; they are about twice as strong as ASS or FSS. *DSS* have significantly better impact toughness and ductility than ferritic *SS*, and show excellent resistance to stress corrosion cracking.

High silicon electrical steels are the low-carbon steels containing up to 4 wt.% silicon. These steels help to reduce eddy current losses; which is beneficial for electrical applications, such as transformer cores and electrical drive components. Depending on the processing technique, silicon electrical steels may be divided into

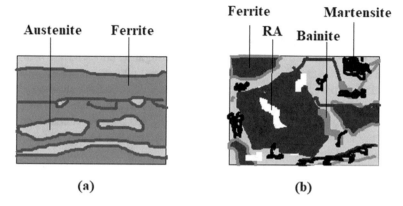

Fig. 5.6 Schematic microstructure of DSS (**a**), and TRIP steel (**b**) (RA = retained austenite)

three types: (a) non-grain oriented electrical steels (NGOES), (b) grain-oriented electrical steels (GOES), and (c) low-carbon electrical steels (LCES). The *GOES* is produced by a complex processing technique, as follows. A hot-rolled coil of thick steel sheet is trimmed, annealed, de-scaled, and cleaned; then the sheet metal is cold rolled via several passes to obtain a sheet thickness in the range of 0.20–0.35 mm (Mazurek, 2012). GOES possess magnetic properties that are strongly oriented along [100] rolling direction (see Chap. 2, Sect. 2.2.5).

TRansformation Induced Plasticity (TRIP) steels are the high alloy steels having high strength and excellent formability. These steels can be used to produce complicated parts. *TRIP steels* possess a multi-phase microstructure, consisting of α-Fe (ferrite), bainite, retained (metastable) γ-Fe (austenite), and martensite (see Fig. 5.6b). TRIP steels possess high energy absorption capacity and fatigue strength, and well-suited for automotive structural and safety parts such as cross-members, longitudinal beams, and bumper reinforcements.

5.6 Non-Ferrous Metals and Alloys

5.6.1 Aluminum and its Alloys

5.6.1.1 Introduction to Aluminum and its Alloys

Aluminum and it alloys are the world's most widely used non-ferrous materials. Commercially pure aluminum has good ductility, corrosion resistance, and electrical conductivity; this is why it find applications in foils and conductor cables. In order to render aluminum suitable for high-strength applications, is necessary to add alloying elements and heat treat it. Owing to their light weight, strength, and corrosion resistance, aluminum alloys find application in aircraft structures, automotive engine components, building construction, and the like (Kaufman, 2000; Huda, 2016).

The strength of an aluminum alloy depends on its alloying contents, degree of cold working, and heat treatment. Aluminum alloys classification systems have been developed for wrought alloys, cast alloys, and heat-treated alloys. Wrought *Al* alloys are strengthened by cold working. The wrought *Al* alloys have a 4-digit system, and the cast alloys have a 3-digit and 1-decimal place system (see Table 5.2). In each designation system, the first digit indicates the principal alloying element. The heat treated *Al* alloys have suffix T3, T-36, T-351, and the like. The series 1xxx *Al* alloys are noted for corrosion resistance; they are used in chemical-handling equipment. The series 6xxx alloys (*e.g.* 6061-T4) are used in automotive engines and pipelines. The cast alloy 356.0 is used in automotive transmission cases and cylinder heads. The heat treatable 2024-T3 and 7075-T6 *Al* alloys are used in aircraft structures, and other stressed applications (see Sect. 5.6.1.2).

5.6.1.2 Aluminum-Silicon Casting Alloys and Aluminum-Copper Aerospace Alloys

Aluminum-Silicon Casting Alloys The metallurgical characteristic of *Al*-Si alloys can be studied by the *Al-Si* phase diagram. The *Al*–12 wt.% *Si* is a good casting alloy since it forms a eutectic at 12.6 wt% Si at 557 °C. Since, the presence of Si in *Al* imparts brittleness, a lower Si content (1.1–7.0 wt% Si) is used in *Al* casting alloys. In industrial practice, small amount of sodium and strontium is added for the refinement of the microstructure of the casting eutectic *Al* alloys (Zamani, 2015). The *Al-Si* casting alloys have tensile strengths in the range of 150–200 MPa, and find applications in automotive engine blocks, car steering-knuckles, and the like.

Table 5.2 The designations systems for cast, heat-treated, and wrought *Al* alloys

Cast alloys series	Principal alloying elements	Heat- treated alloys suffix codes	Heat treatment, temper, and post-process	Wrought Alloys series	Principal alloying elements
1xx.x	99%*Al*	xxx-T3	Solution heat treated (SHT), then cold worked (CW)	1xxx	99.xx%*Al*
2xx.x	Copper	xxxx-T36	SHT, and CW (controlled)	2xxx	Copper
3xx.x	Si + Cu and/or Mg	xxxx-T351	SHT, stress-relieved stretched (SRS), then CW	3xxx	Manganese
4xx.x	Silicon	xxxx-T4	SHT, then naturally aged	4xxx	Silicon
5xx.x	Magnesium	xxxx-T451	SHT, then SRS	5xxx	Magnesium
6xx.x	Unused series	xxxx-T5	Artificially aged only	6xxx	Mg + Si
7xx.x	Zinc	xxxx-T6	SHT, artificial aging	7xxx	Zinc
8xx.x	Other elements	xxxx-T6	SHT, then artificially aged	8xxx	Other elements (*e.g.* Li)

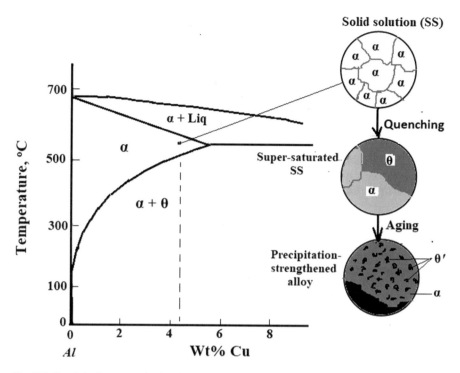

Fig. 5.7 Precipitation strengthening heat treatment of *Al-Cu* alloys

Aluminum-Copper Aerospace Alloys *Aluminum-copper alloys* are strong and light-weight; they are noted for aerospace applications. The heat-treatable 7075-T6 *Al* alloys has yield strength high than 500 MPa, and finds applications in aircraft fittings, gears and shafts, missile components, regulating valves, and the like. The series 2xxx *Al* alloys (containing 1.9–6.8% Cu and some Mn, Mg, and Zn) are used in forgings, extruded parts, and liquefied-gas storage tanks in commercial and military aircrafts. In particular, the age-hardenable 2024-T4 *Al-Cu* alloy is given a precipitation-strengthening heat-treatment process for aerospace application. In this heat-treatment process, an *Al*–4.2 wt.% Cu alloy (say) is selected, and given a solution treatment by heating in the range of 450–550 °C and then holding at the temperature for 1 h followed by quenching so as to obtain a super-saturated solid solution. The quenched alloy is then aged by heating at about 160 °C; this treatment finally produces fine dispersed θ' phase in α solid-solution (see Fig. 5.7). The tensile strength of the precipitation-strengthened *Al* alloy (2024-T4) is over 450 MPa.

Another example of aerospace aluminum alloy is alloy 2024-T81; which has a high tensile yield strength and ultimate tensile strength of 450 MPa and 485 MPa, respectively. In designing aerospace alloys, for tensile loading below the yield limit, the applied stress on a structural component should be considered for its specific yield strength (yield strength per unit density). In comparing specific strength of two aerospace materials, the design relationship is expressed as:

$$\frac{S_a}{S_b} = \frac{\sigma_{ys(a)} \cdot \rho_b}{\sigma_{ys(b)} \cdot \rho_a} \tag{5.3}$$

where S_a is the specific strength of the part using material (a), S_b is the specific strength of the part using material (b); ρ_a and ρ_b are the densities of materials (a) and (b), respectively; and $\sigma_{ys(a)}$ and $\sigma_{ys(b)}$ are the yield strengths of materials (a) and (b), respectively (see Example 5.6).

5.6.2   Copper and its Alloys

Copper is an excellent electrical conductor, and hence finds wide applications in electrical industry. Notable examples of copper alloys include: brasses (copper-zinc), bronzes (copper-tin), gun metal (Cu-Sn-Zn-Pb), Cu-Ni alloys, nickel silver (Cu-Ni-Zn), and the like. Copper alloys for high temperature applications (*e.g.* in boilers) include: silicon bronzes, aluminum brasses, and copper nickels. Copper and its alloys find applications in electrical industry, architecture, automotive, telecommunication, and mechanical engineering (tubes, marine, pipe, and fittings, fuel gas piping system, etc.). Copper tubes are used as refrigerant lines in HVAC systems.

Brasses are very important engineering materials; these include red brass, cartridge brass, clock brass, and *Muntz metal*. Alpha (α) brasses contain up to 38 wt.% Zn, and are relatively ductile with good formability; and find applications in deep drawn and cold-worked components (e.g. automotive radiator caps, battery terminals, etc.). Cartridge brass is 70%Cu 30%Zn alloy. Additionally, there are other phases (β, γ, ε, η) that are stable at various temperatures corresponding to various compositions of Cu-Zn alloys. The β-brasses are suitable for highly-stressed applications, but they must be hot-worked because of their restricted ductility below 450 °C (if cold worked). Cold working significantly improves the strength of brass. The tensile strength of cold-worked cartridge brass wire may be as high as 900 MPa. Some alloying elements (lead, arsenic, nickel, iron, etc.) are added to the basic copper-zinc alloys for: (a) improving tensile strength and wear resistance, (b) improving machinability, and (c) improving corrosion resistance.

Bronzes (or copper-tin alloys) are known for their remarkable corrosion resistance. They have higher tensile strength and ductility than red brass and semi-red brass. Bronzes, with up to 15.8% Sn, retain the FCC structure of alpha copper. Bronzes are used in bearings, gears, piston rings, valves and fittings. Lead is added to bronzes to improve machinability for application in sleeve bearings. Since lead decreases the tensile strength of the bronzes, the composition is usually adjusted to balance machinability and strength requirements.

5.6.3 Nickel and its Alloys

Nickel has a low coefficient of thermal expansion; and finds application in bimetallic thermocouple sensors. Nickel alloys have excellent tensile strength and resistances to creep and hot corrosion resistance *Monel metal* is a nickel-copper alloy with excellent resistance to corrosion. *Monel 400* (63% Ni, 30% Cu, 2.3% Fe, and 1.8% Mn) is one of the few alloys that maintains its strength at sub-zero temperatures. The most important group of nickel alloys, is the superalloys; which possess high tensile and creep strength with high resistance to hot corrosion. The chief alloying elements in nickel-base superalloys include: Cr, Fe, Mo, W, Ti, Al, C, Ta, Zr, Nb, B, and the like. Ni-base superalloys find extensive applications in hot-section components (*e.g.* vanes, disks, and blades) of gas-turbine (GT) engines and nuclear reactors. Some nickel-base superalloys developed for application in the GT blades include: *GTD-111*, Allvac 718Plus, and the single crystal superalloys (*e.g.* Rene-N6 and MC-534) (Huda, 2017).

The creep behavior of Ni-base superalloys is crucial to aerospace and nuclear industries. The precipitation-strengthened superalloys with fine dispersed γ' precipitates with a coarse grain size of γ (FFC) nickel matrix in the microstructure ensures a good creep strength (Huda, 2007). It has been experimentally proved that at a high temperature of 850 °C, the creep strain rates in a coarse-grained and a fine-grained superalloy at 250 MPa may be co-related by (Thébaud, et al., 2018):

$$\dot{\varepsilon}_{(FG)} = 100 * \dot{\varepsilon}_{(CG)} \tag{5.4}$$

where $\dot{\varepsilon}_{(FG)}$ and $\dot{\varepsilon}_{(CG)}$ are the creep strain rates in a fine grained (FG) and the coarse grained (CG) superalloy, respectively (see Example 5.7).

5.6.4 Titanium and its Alloys

Commercially purity (99 − 99.5%) titanium has a low tensile strength ($\sigma_{ut} =$ 330–650 MPa); however, when titanium is alloyed with *Al*, Cu, Mn, Mo, Sn, V, or Zr, its strength significantly increases to 80–1450 MPa. Titanium alloys have a good high-temperature strength. The stiffness (Young's modulus) of titanium alloys is excellent ($E = 125$ GPa for the Ti-6Al-4 V alloy). The fatigue endurance strength at 10^7 cycles for titanium alloys is about 2/third of tensile strength. In addition to these exceptional mechanical properties, titanium alloys have a low density of about 4500 kg/m³; which is slightly over half that of steel. Owing to their excellent specific strength combined with corrosion resistance, titanium alloys are extensively used as in aircraft and spacecraft structures, gas turbine components (e.g. compressor blades), guided missiles, and the like. Additionally, a good bio-compatibility of titanium alloys allows us to use them as biomaterials in artificial hips and other orthopedic bioengineering applications.

5.7   Ceramics and Glasses

5.7.1   Introduction to Ceramics

Ceramics are non-organic solids having high melting temperatures; they may be either crystalline or amorphous (e.g. glass). They are hard and brittle. Although their tensile strength is low but compressive strength is high. They have high creep resistance. Examples of ceramics include: oxides (*e.g.* alumina, silica, zirconia, etc), carbides (*e.g.* silicon carbide, boron carbide, etc), nitrides (*e.g.* silicon nitride, boron nitride, etc), and diamond. Glasses and ceramics may be broadly classified into two groups: (a) traditional ceramics, and (b) advanced ceramics.

5.7.2   Traditional Ceramics

Traditional ceramics are made up of three basic components: (a) clay, (b) silica (quartz), and (c) feldspar. Clay is primarily composed of hydrated aluminum silicates ($Al_2O_3.SiO_2.H_2O$). It is one of the most common ceramic raw materials; and is widely used in structural clay products (bricks, pipes, tiles, etc) and white-wares (pottery, table-ware, sanitary-ware, etc). Silica bricks are used as furnace refractories, which involves a great industrial importance since any failure in refractories would cause heavy damages. Refractories must, therefore, resist aggressive furnace conditions, including high temperatures, chemical and acid attack, abrasion, mechanical impact, and the like. Glass (a super-cooled silica) is an important ceramic material in our society since it is extensively used in construction/automotive industry as well as in other engineering applications. Building construction industry is the largest user of traditional ceramics; for example ceramic tiles are used in floor, bath-room, and kitchen construction (see Fig. 5.8).

Fig. 5.8 Ceramic tiles in floor and kitchen construction

Table 5.3 Thermal properties of some ceramic materials

Material	Melting temperature (°C)	Thermal conductivity (W/m K)	Heat capacity (J/kg K)	Coefficient of linear expansion (1/°C)
Soda-lime glass	700	1.7	840	9.0×10^{-6}
Fused silica	1650	2.0	740	0.5×10^{-6}
Alumina	2050	30.1	775	8.8×10^{-6}

5.7.3   Advanced Ceramics

Advanced ceramics (or engineering ceramics) are special types of ceramics used mainly for electrical, electronic, optical, and magnetic applications. Special properties (*e.g.* ferro-electricity) of advanced ceramic materials (*e.g.* lead zirconate titanate, lead titanate, barium titanate, etc) are utilized for their industrial applications, such as sensors, pumps, sonar, microphones, and the like. Magnetic ceramics are used for the production of antennas and inductors. Bio-ceramics (such as alumina) with high density and purity are used in dental implants, eye glasses, chemical ware, and the replacement of hips and knees.

Structural applications of advanced ceramics usually take advantage their superior thermal properties; some of these applications include: components of automobile engines (*e.g.* spark plug body), armor for military vehicles, and space shuttle structures. Some thermal properties of common ceramic materials are presented in Table 5.3. The application of the ceramics in space shuttles makes use of their high melting temperatures. The exterior of space shuttle is covered with ceramic tiles made from high purity amorphous silica fibers. The shuttle's parts exposed to the highest temperatures have an added layer of high-emittance glass. These ceramic tiles can withstand temperatures up to 1480 °C for a specified duration of time.

5.8   Polymers and Plastics

5.8.1   Introduction to Polymers and Plastics

Polymers are organic materials (hydrocarbons) that come from repeating molecules or macromolecules. The building blocks for making polymers are small *organic* molecules - molecules that contain carbon along with other substances. They generally come from oil (petroleum) or natural gas. Each of these small molecules is known as a *monomer*; which is capable of joining with other monomers to form very long molecule chains called *polymers*. This process is called *polymerization*. For example, many (*n* number of) ethylene molecules (monomers) may unite to form polyethylene (polymer) according to the following polymerization reaction:

_n_ ethylene monomers   **Polyethylene (polymer)**

The average number of repeat units in a polymeric chain is called the *degree of polymerization* (DP); which is the ratio of number-average molecular weight to the repeat unit molecular weight.

$$DP = \frac{Number\ average\ molecular\ weight}{Repeat\ unit\ molecular\ weight} = \frac{M}{m} \tag{5.6}$$

The significance of Eq. 5.6 is illustrated in Example 5.8.

A polymeric material may be partially amorphous and partially crystalline. The degree of crystallinity by weight, for a polymer, may be calculated by:

$$X_c = \frac{(\rho - \rho_a)\rho_c}{(\rho_c - \rho_a)\rho} \tag{5.7}$$

where X_c is the weight fraction of crystalline material in the polymer sample; ρ is the total density of the sample; ρ_a is the density of the fully amorphous polymer; and ρ_c is the density of the fully crystalline polymer (see Example 5.9).

Plastics are synthetic polymers; they are so named because plastics can be molded and shaped for a wide variety of products. Plastics are classified into two categories according to what happens to them when they are heated to high temperatures. *Thermoplastics* keep their plastic properties: They melt when heated, and then harden again when cooled. *Thermosets,* on the other hand, are permanently "set" once they are initially formed and cannot be melted. If they are exposed to enough heat, they would crack or become charred.

5.8.2 *Plastics* – **Mechanical Behaviors and Applications**

There are many different types of plastic material with useful and versatile properties. Some commonly used plastics include: poly-vinyl chloride (PVC), polyethylene terephthalate (PET), high density poly-ethylene (HDPE), low density poly-ethylene (LDPE), and polypropylene (PP). The mechanical behaviors and uses of these types of plastic are briefly explained as follows.

Polyvinyl Chloride (PVC)   *PVC* is a low-cost thermoplastic; it is highly resistant to chemical and biological damage. PVC is easy to fabricate and mold into shapes; it can be used to create rigid, lightweight sheets, like Foamex. PVC find applications in signage, clothing, furniture, medical containers, tubing, water bottles, water and sewage pipes, cables, and the like.

Polyethylene Terephthalate (PET)   *PET* can be completely rigid or flexible. Owing to its molecular construction, *PET* has high impact toughness. It is resistant to chemical and weather attack. Common applications of *PET* include: bottles of water, soft drink, and cooking oil; first-aid blankets; packaging trays; frozen ready-meal; and the like.

High Density Polyethylene (HDPE)   *HDPE* is a thermoplastic, and has good strength due to its high density. It is capable of withstanding high temperatures and strong chemicals. *HDPE* is bio-degradable *i.e.* it can be recycled in many different ways; and therefore it can be converted into many different products. Common uses of *HDPE* include: shopping bags; containers for cleaning solution, soap, and food/drink storage; pipes; bottle caps; vehicle fuel tanks; and the like.

Low Density Polyethylene (LDPE)   At moderate temperatures, *LDPE* is a highly non-reactive thermoplastic material; this is why it is one of the most common plastics. It can withstand temperatures close to 100 °C. It has high resiliency. *LDPE* finds applications in containers, trays, machine parts, lids, drink cartons, protective shells, computer casings, bin-bags, laundry bags, etc.

Polypropylene (PP)   *PP* is a thermosetting plastic; it is strong and flexible. When melted, it is one of the most effective materials for injection molding. Common applications of PP include: bottle caps, clothing, surgery tools, food containers, straws, lunch boxes, and the like.

Epoxy Resins   *Epoxy* is a thermosetting resin. The term "epoxy" refers to a broad group of reactive compounds that are characterized by the presence of an oxirane or epoxy ring. *Epoxy* is represented by a three-member ring containing an oxygen atom that is bonded with two carbon atoms already united in some other way. Epoxy resins possess high strength ($\sigma_{ut} = 90–120$ MPa), high resistance to chemical attack, low shrinkage, excellent adhesion to various substrates, effective electrical insulation, low cost and low toxicity. Epoxy resins find wide applications in adhesives; paints and coatings; composite materials (*e.g.* in CFRP), tooling, electrical systems and electronics, consumer products, and the applications in marine-, aerospace-, and bio-engineering.

5.9   Composite Materials

A composite material is a combination of two materials with different properties. The component materials work together to give the composite unique properties, but within the composite, the components materials do not dissolve or blend into each other. Composites may be classified based on two classification systems: (a) based on the matrix material (metal, ceramic, polymer) and (b) based on the reinforcing material structure (see Chap. 15, Fig. 15.1). *Metal matrix composites (MMCs)* are composed of a metallic matrix (*e.g.* aluminum, magnesium, etc.) and a dispersed ceramic or metallic phase. *Ceramic matrix composites (CMCs)* are composed of a ceramic matrix and embedded fibers of other ceramic material as dispersed phase. *Polymer matrix composites (PMCs)* are composed of a polymer matrix (a thermoplastic or thermoset) and a dispersed phase (*e.g.* (embedded glass, carbon, steel or Kevlar fibers). As regards the types of composite based on the reinforcing material structure, there are three types of composites: (1) particulate composites, (2) fibrous composites, and (3) laminate composites. *Particulate composites* consist of a matrix reinforced by a dispersed phase in the form of particles (see Fig. 5.9). *Fiber reinforced composites* consist of a matrix reinforced by an either dispersed phase (discontinuous fibers) or aligned/continuous fibers. A laminate composite is composed of 2-dimensional sheets or panels that have a preferred high strength in a particular direction.

Composite materials find wide range of applications in the following industry sectors: aerospace, automotive, construction, marine, bio-medical, defense, consumer products, machine components, and the like. Cemented carbide is commonly used as a cutting tool in machining. Currently, advanced composite materials are being increasingly used in aerospace industry. The composite materials offer high specific strength (strength-to-weight ratio), corrosion resistance, and high fatigue resistance. In particular, carbon fiber reinforced polymer (CFRP) composites, possessing the best specific strength, are being used in both military and passenger aircrafts (see Chap. 15, Table 15.1). Aircrafts, rockets, and missiles all fly higher with the use of composites in their structures. Mechanical behavior of composites is explained in Chap. 15.

Fig. 5.9 Schematic illustration of a particulate composite

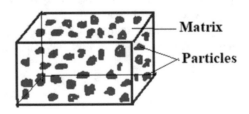

5.10   Semiconductors and Advanced Materials

Semiconductors   *Semiconductors* are the electronic materials with electrical conductivities intermediate between those of conductors and insulators. They are the backbone of today's electronic industry. The three semiconducting elements (silicon, germanium, and tin), from column IVA of the periodic table, serve as a kind of boundary between metallic and nonmetallic elements. Semiconductors may be classified into two main types: (1) intrinsic semiconductors, and (2) extrinsic semiconductors (see Fig. 1.1). The scope of this does not allow this author to discuss semiconductors in detail; the reader is therefore advised to refer to literature (*e.g.* Barbero, 2018).

Advanced Materials   *Advanced materials* are the materials that are specifically engineered to exhibit novel or enhanced properties that result in superior performance relative to conventional materials. Examples of advanced materials include: nanomaterials, smart materials, and the like. Nanomaterials are those materials with feature size (particle diameter, grain size, layer thickness, etc) smaller than 100 nm. The scope of this does not permit this author to discuss semiconductors in detail; the reader is therefore advised to refer to literature (*e.g.* Ven et al. 2020).

5.11   Calculations – *Worked Examples/Solved Problems*

Example 5.1 Calculating the Carbon Equivalent and Identifying Cast Irons   A samples of cast iron (*CI*) is provided to you. The composition of the sample is: 2.4 wt.% carbon, 1.2 wt.% silicon, 0.14 wt.% phosphorous, 0.4 wt.% manganese, balance (wt%) iron. Calculate the carbon equivalent for the *CI* sample. Is this a gray *CI* sample or white *CI* sample?

Solution   C = 2.4, Si = 1.2, P = 0.14, CE = ?

By using Eq. 5.1,

$$CE\left(\text{in wt}\%\right) = C + \frac{Si + P}{3} = 2.4 + \frac{1.2 + 0.14}{3} = 2.84$$

Since the value CE < 4.3, the sample is white cast iron.

Example 5.2 Calculating the Nodules Count Per Unit Volume for SG Cast Iron   The nodules count per unit area for a sample of SG cast iron is 110 nod/mm². Calculate the graphite nodule count per unit volume for the SG iron. Given: the parameter $\alpha = 1.3$, and the fraction of graphite, f_g is 0.1.

Solution $N_A = 110$ nod/mm², $f_g = 0.1$, $\alpha = 1.3$

By using Eq. 5.2,

$$N_V = \sqrt{\frac{\pi}{6f_g}}\left(\alpha\,N_A\right)^{3/2} = \sqrt{\frac{\pi}{6\times0.1}}\left(1.3\times110\right)^{1.5} = 2.288\times1710 = 3913\,\text{nod}\,/\,\text{mm}^3$$

The graphite nodules count per unit volume = 4925 nod/mm^3

Example 5.3 Determining the Mechanical Properties of a Low-carbon Steel from Plot By reference to Fig. 5.5, determine the tensile strength (σ_{ut}), yield strength (σ_{ys}), and % reduction in area for a low-carbon steel. Is this steel suitable for use in cutting tools?

Solution By reference to Fig. 5.5, a low-carbon steel containing 0.1 wt.% C possesses the following mechanical properties: σ_{ut} = 400 MPa, σ_{ys} = 260 MPa, % reduction in area = 75.

The steel is unsuitable for use in cutting tools since the strength or hardness is too low.

Example 5.4 Determining the Mechanical Properties of a Medium-carbon Steel from Plot By reference to Fig. 5.5, determine the tensile strength (σ_{ut}), yield strength (σ_{ys}), and % reduction in area for an eutectoid steel. Is this steel suitable for moderate stressed applications?

Solution (a) By reference to Fig. 5.5, the eutectoid (0.77 wt.% C) steel possess the following properties:

σ_{ut} = 770 MPa, σ_{ys} = 350 MPa, % reduction in area = 30.

Yes! The steel, after proper heat treatment, is suitable for moderate stressed applications.

Example 5.5 Determining Composition of Carbon Steel with Best Mechanical Properties By reference to Fig. 5.5, determine the composition of annealed carbon steel corresponding to: (a) the best combination of σ_{ys} and ductility, and (b) the best combination of σ_{ut} and ductility.

Solution By reference to Fig. 5.5,

(a) The composition of annealed carbon steel corresponding to the best combination of yield strength and ductility in annealed carbon steel is Fe–0.3 wt.% C.

(b) The composition of annealed carbon steel corresponding to the best combination of tensile strength and ductility is Fe–0.14 wt.% C.

Example 5.6 Comparing Specific Strengths of Two Aerospace Alloys The densities of 7075-T6 *Al* alloy and aluminum are 2.81 and 2.7 g/cm^3, respectively. The yield strengths of the same materials are 503 MPa and 240 MPa, respectively. Compare the specific strengths of the two materials, and select the appropriate material for aerospace application.

Solution $\rho_a = 2.81$ g/cm^3, $\rho_b = 2.7$ g/cm^3, $\sigma_{ys(a)} = 503$ MPa, $\sigma_{ys(b)} = 240$ MPa, $\dfrac{S_a}{S_b} = ?$

By using Eq. 5.3,

$$\text{Ratio of the specific strengths} = \frac{S_a}{S_b} = \frac{\sigma_{ys(a)} \cdot \rho_b}{\sigma_{ys(b)} \cdot \rho_a} = \frac{503 \times 2.7}{240 \times 2.81} = 2$$

Since the 2024-T81 *Al* alloy has double the specific strength compared to aluminum, the former is appropriate for aerospace structural application.

Example 5.7 Calculating the Creep Strain rate in a Fine-Grained Ni-base Superalloy A creep strain rate of 5.2×10^{-9} /s was observed at 850°C/250MPa for a nickel-base superalloy with a grain size of 370 μm. Calculate the creep strain rate in the superalloy with a grain size of 50 μm at the same testing conditions.

Solution $\dot{\varepsilon}_{(CG)} = 5.2 \times 10^{-9}$ /s, $\dot{\varepsilon}_{(FG)} = ?$

By using Eq. 5.4,

$$\dot{\varepsilon}_{(FG)} = 100 * \dot{\varepsilon}_{(CG)} = 100 \times 5.2 \times 10^{-9} = 5.2 \times 10^{-7} \text{ /s}$$

For fine-grained superalloy, the creep strain rate increases to 5.2×10^{-7}/s.

Example 5.8 Calculating the Degree of Polymerization of Poly-propylene (PP) The repeat unit of poly-propylene (PP) is shown in Fig. 5.10. The number-average molecular weight of PP is 33,040 g/mol. Calculate the degree of polymerization of PP.

Solution The repeat unit molecular weight of PP $= m = (3 \times$ at. wt. of C$) + (6$ at. wt. of H$)$

$$m = (3 \times 12.01) + (6 \times 1.008) = 42.08 \, g/mol$$

By using Eq. 5.6,

$$DP = \frac{M}{m} = \frac{33,040}{42.08} = 785$$

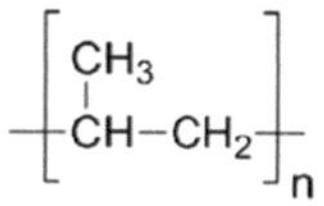

Fig. 5.10 Repeat unit of
poly-propylene (PP)

Example 5.9 Calculating the Weight Fraction and % Crystallinity in a Polymer Sample The density of the fully crystalline polyethylene is 0.998 g/cm³, and that of fully amorphous polyethylene is 0.870 g/cm³. The total density of a polyethylene sample is 0.930 g/cm³. Calculate the (a) weight fraction of the crystalline material in the polymer sample, (b) % crystallinity.

Solution $\rho = 0.930$ g/cm³, $\rho_a = 0.870$ g/cm³, $\rho_c = 0.998$ g/cm³, $X_c = ?$

By using Eq. 5.7,

$$X_c = \frac{(\rho - \rho_a)\rho_c}{(\rho_c - \rho_a)\rho} = \frac{(0.93 - 0.87)\times 0.998}{(0.998 - 0.870)\times 0.93} = \frac{0.0598}{0.1190} = 0.502$$

The weight fraction of the crystalline material in the polyethylene sample = 0.502
%crystallinity = $X_c \times 100 = 50.2$

Questions and Problems

5.1. Refer to Table 5.2. (a) What is the principal alloying element in the 7075-T6 aluminum alloy? (b) What heat treatment is given to the alloy? (c) List three applications of this alloy.
5.2. What is the meaning of each of the following acronyms used in metals and alloys?
        (a) 2024-T4, (b) DSS, (c) ADI, (d) AISI, (e) SAE, (f) CE, (g) TRIP steel, (h) GOES.
5.3. Differentiate between the four types of carbon steels.
5.4. Justify the applications of alloy steels in automotive and highly stressed applications.
5.5. Why is a small amount of sodium or strontium added to the eutectic *Al-Si* alloy?
5.6. What heat treatment process is given to 2024-T4 aluminum alloys? Explain the process.
5.7. Discuss the characteristics and applications of alpha (α) and beta (β) brasses.
5.8. What are bronzes? List the general characteristics and applications of bronzes.
5.9. List three applications of nickel and its alloys, and justify them.
5.10. Justify the aerospace and bio-medical applications of titanium alloys.
5.11. (a) Distinguish between thermoplastics and thermosetting plastics. Give their examples.
        (b) Explain the characteristics and mechanical behavior of epoxy resins.
5.12. Highlight the characteristics and applications of traditional and advanced ceramics.
5.13. (a) Classify composites based on: (a) matrix material, and (b) reinforced material structure.

(b) Which composite material is the most extensively used in aircrafts? And why?

5.14. Calculate the carbon equivalent of the cast iron having composition: Fe–2.6%C–1.3%Si– 0.17%P–0.5%Mn.

5.15. The densities of the 2024-T81 *Al* alloy and carbon steel are 2.87 and 7.87 g/cm^3, respectively. The yield strengths of the same alloys are 515 MPa and 483 MPa, respectively. Compare the specific strengths of the two alloys.

5.16. A creep strain rate of 8.6 x 10^{-9}/s was observed at 850 °C/250 MPa for a nickel-base superalloy with a grain size of 300 μm. Calculate the creep strain rate in the superalloy with a grain size of 70 μm at the same testing conditions.

5.17. (MCQs). Encircle the most appropriate answers for the following statements:

(a) Which steel is the most suitable for use in cutting tools?
    (i) High-speed steel, (ii) low carbon steel, (iii) medium carbon steel, (iv) high carbon steel.

(b) Which steel has the highest ductility and the lowest hardness?
    (i) low carbon steel, (ii) medium carbon steel, (iii) mild steel, (iv) high carbon steel.

(c) Which steel has the best combination of strength and ductility?
    (i) low carbon steel, (ii) medium carbon steel, (iii) high carbon steel, (iv) MS.

(d) Which type of LAHS steel is suitable for applications in axles, shafts, and gears?
    (i) Cr-W-V, (ii) Ni-Cr-Mo, (iii) Mn-Cr-C, (iv) High nickel

(e) Which alloy steel's microstructure has ferrite, bainite, martensite, and retained austenite?
    (i) TRIP steel, (ii) MSS, (iii) DSS, (iv) Hadfield Mn steel

(f) Which cast iron has the best fatigue strength and damping capacity?
    (i) AD iron, (ii) SG iron (iii) CG iron, (iv) Alloyed CI

(g) Which type of cast iron has the best fracture toughness?
    (i) AD iron, (ii) SG iron (iii) CG iron, (iv) Alloyed CI

(h) Which alloy is the best bio-material?
    (i) steel, (ii) copper alloy (iii) titanium alloy, (iv) aluminum alloy

(i) Which material is the most extensively used in hot sections of gas turbine engines?
    (i) composites, (ii) polymers, (iii) superalloy, (iv) titanium alloy

(j) Which alloy is used in the compressor blades of aircraft engines?
    (i) aluminum alloy, (ii) superalloy, (iii) copper alloy, (iv) titanium alloy

(k) Which material has the best specific strength (strength to weight ratio)?
    (i) CFRP, (ii) GFRP (iii) aluminum, (iv) steel

(l) Which metal is the most commonly used in electrical industry?
    (i) aluminum, (ii) zinc, (iii) copper, (iv) titanium

(m) Which designation of *Al* alloys refers to solution heat treated, then artificial aged?
    (i) xxxx-T6, (ii) xxxx-T4, (iii) xxxx-T5, (iv) xxxx-T36

(n) Which designation series of *Al* alloys refers to copper as the main alloying element?
   (i) 2xxx, (ii) 4xxx, (iii) 5xxx, (iv) 7xxx

(o) Which phase in Ni-base superalloy's microstructure mainly imparts
creep strength?
   (i) carbide, γ, (ii) gamma prime, γ' (iii) sigma, σ, (iv) gamma

(p) Which type of plastic is the lowest cost thermoplastic?
   (i) PP, (ii) PVC, (iii) HDPE, (iv) LDPE

(q) Which ferro-electric material is used in electronic sensors?
   (i) barium titanate, (ii) aluminum, (iii) HDPE, (iv) copper.

References

Barbero EJ (2018) Introduction to composite materials design. CRC Press, Boca Raton

Huda Z (2020) Metallurgy for physicists and Engineers. CRC Press, Boca Raton

Huda Z (2007) Development of heat treatment process for a P/M superalloy for gas-turbine blades.
   Mater Des 28(5):1664–1667

Huda Z (2017) Energy-efficient gas-turbine blade-material technology – *A review*. Mater Technol
   3(51):355–361

Huda Z (2016) Materials selection in design of structures of subsonic and supersonic aircrafts. In:
   Liu Y (ed) *Frontiers in aerospace science*, vol 1, pp 457–481

Jones AL (2020) Modern cast irons - the complete guide to selecting, seasoning, cooking, and
   more, Red Lightning Books. Indiana University Press, Bloomington

Kaufman JG (2000) Introduction to aluminum alloys and tempers. ASM International,
   Materials Park

Thébaud L, Villechaise P, Coraline C, Cormier J (2018) Is there an optimal grain size for creep
   resistance in Ni-based disk superalloys? Mater Sci Eng A 716:274–283

Mazurek, R. (2012). Effects of burrs on a three phase transformer core including local loss, total
   loss and flux distribution. *PhD Thesis*, Cardiff School of Engineering, Cardiff University, UK

Ven v d, Gauvin R, Soldera A (2020) Advanced materials. De Gruyter, Berlin

Zamani, M. (2015). Al-Si cast alloys – microstructure and mechanical properties at ambient and
   elevated temperatures. Licentiate Thesis. School of Engineering, Jönköping University, Sweden

Part II
Stresses, Strains, and Deformation Behaviors

Chapter 6
Stress-Strain Relations and Deformation Models.

6.1 True Stress and True Strain

We, in Chap. 3, established mathematical expressions for engineering stress and strain. In engineering design and analysis, equations describing stress–strain behavior, or stress–strain relationships, are frequently needed. In this section, we will therefore establish relationships between stresses and strains by introducing true stress and true strain. In order to accurately specify the plastic behavior of ductile materials by considering the actual (instantaneous) dimensions, true stress (σ_{true}) and true strain (ε_{true}) are used; which are defined as follows (see Fig. 6.1a):

$$\sigma_{true} = \frac{F}{A} \qquad (6.1)$$

where F is the tensile force, and A is the cross-sectional area at any instant during deformation.

$$\varepsilon_{true} = \int_{L_0}^{L} \frac{dL}{L} = \ln\left(\frac{L}{L_0}\right) \qquad (6.2)$$

where A is the cross-sectional area, mm²; L_0 is the original gage length, mm; and L is the length at any instant during the tensile loading, mm (see *Example 6.1*).

By reference to Fig. 6.1(a) and by using the constant volume relationship, we obtain:

$$A \cdot L = A_0 \cdot L_0 \qquad (6.3)$$

By combining Eq. 6.1 and 6.3,

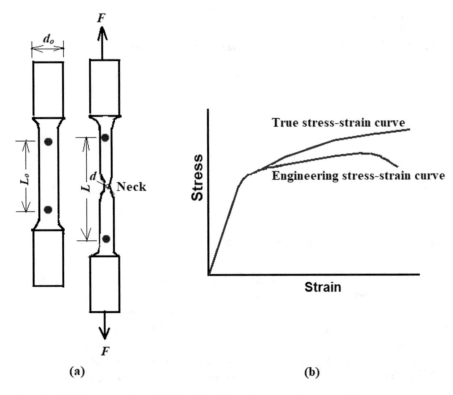

Fig. 6.1 Tensile test specimen under loading (**a**), and stress-strain curves (**b**)

$$\sigma_{true} = \frac{F}{A} = \frac{F * L}{A_0 * L_0} \tag{6.4}$$

The significance of Eq. 6.4 is illustrated in *Example 6.2*.

By combining Eqs. 3.5, 3.6, and 6.4, we get:

$$\sigma_{true} = \sigma_{eng}\left(1 + \varepsilon_{eng}\right) \tag{6.5}$$

By combining Eq. 3.6 and 6.2, we obtain:

$$\varepsilon_{true} = \ln\left(1 + \varepsilon_{eng}\right) \tag{6.6}$$

Eqs. 6.5 and 6.6 can be used to draw the true stress-strain curve from the engineering stress-strain curve, as shown in Fig. 6.1(b) (see also *Example 6.3*).

By combining Eq. 3.6, 6.3, and 6.6, we get:

$$\varepsilon_{true} = \ln\left(1 + \varepsilon_{eng}\right) = \ln\left(\frac{A_0}{A}\right) \tag{6.7}$$

6.2   Stress-Strain Relationships – *Young's–*, *Tangent–*, and *Plastic Moduli*

We learnt in Chap. 3 that the tensile testing of a material involves two main deformations – elastic deformation and plastic deformation (see Fig. 3.5 & Fig. 6.2).

In order to establish a useful stress-strain relationship that encompasses both elastic and plastic strains, it is technologically important to introduce three moduli – the Young's modulus (*E*), the *plastic modulus* (*H*), and the *tangent modulus* (*K*) (Bertram and Glüge 2015). The Young's modulus (*E*) is the slope of the stress-strain curve in the elastic region *i.e.*

$$E = \frac{d\sigma}{d\varepsilon_e} \tag{6.8}$$

where *dσ* is the increment in stress and $d\varepsilon_e$ is the increment in elastic strain (see Fig. 6.2). Similarly, the plastic modulus (*H*) is the slope of the stress-strain curve in the plastic region *i.e.*

$$H = \frac{d\sigma}{d\varepsilon_p} \tag{6.9}$$

where $d\varepsilon_p$ is the increment in plastic strain (see Fig. 6.2). At any instant of strain, the increment in stress (*dσ*) is related to the overall increment in strain (*dε*), by:

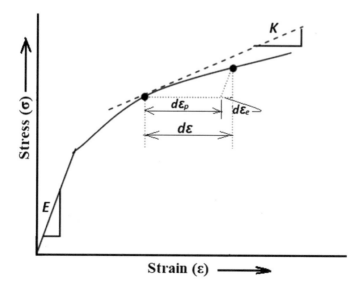

Fig. 6.2 Stress-strain curve showing elastic strain change ($d\varepsilon_e$) and plastic strain change ($d\varepsilon_p$)

Table 6.1 The data for n and K for some alloys (see Eq. 6.13)

Alloy	Copper (annealed)	Naval brass (annealed)	Low carbon steel (annealed)	304 stainless steel	4340 alloy steel	2024-T3 Al alloy
N	0.44	0.21	0.21	0.44	0.12	0.17
K (MPa)	530	585	600	1400	2650	780

$$K = \frac{d\sigma}{d\varepsilon} \tag{6.10}$$

where K is the tangent modulus or the strength coefficient of the material (see Table 6.1).

It is evident in Fig. 6.2 that the increment in strain ($d\varepsilon$) is related to the increment in elastic strain ($d\varepsilon_e$) and the increment in plastic strain ($d\varepsilon_p$) by:

$$d\varepsilon = d\varepsilon_e + d\varepsilon_p \tag{6.11}$$

By combining Eqs. 6.8, 6.9, 6.10, and 6.11, we obtain:

$$\text{or} \qquad \frac{1}{K} = \frac{1}{E} + \frac{1}{H} \tag{6.12}$$

The significance of Eqs. 6.8, 6.9, 6.10, 6.11, and 6.12 is illustrated in *Examples 6.4 and 6.5*.

6.3 Stress-Strain Relationship in Strain Hardening

We learnt in Chap. 4 (Sect. 4.3) that many work-hardened metals and alloys have stress-strain curves; which can be represented by the following power-hardening stress–strain relationship, (Huda 2020; Huda and Bulpett 2012):

$$\sigma_{true} = K\varepsilon_{true}^{n} \tag{6.13}$$

where σ_{true} is the true stress (or the yield stress); ε_{true} is the true strain; K is the strength coefficient (or the tangent modulus); and n is the work-hardening exponent (see Examples 6.6, 6.7, and 6.8). For a perfectly plastic solid, $n = 0$; whilst $n = 1$ represents a 100% elastic solid. A metal (or alloy) can be strengthened by strain hardening by selecting an n value in the range of 0.10–0.50. The data for n and K for various alloys are presented in Table 6.1.

6.4   Elastic and Plastic Deformation Modles – *Yield Criteria*

We have been introduced to elastic and plastic deformations in Chap. 1. It may be recalled that *elastic* deformation is associated with the stretching/compression, but not breaking, of chemical bonds. On the other hand, in plastic deformation, atoms change their relative positions. Elastic deformations are reversible *i.e.* the energy expended in deformation is stored as elastic strain energy and is completely recovered upon the removal of load. In general, almost all engineering materials undergo some permanent (plastic) deformation, which remains after the removal of load. Most engineering materials, with a few exception, exhibit both elastic and plastic deformation behaviors. However, there are some exceptions to these deformation behaviors; for example in some metal forming operations involving large plastic deformation, the elastic deformation may be ignored. The various deformation behaviors may be represented by four simple deformation models: (a) linear elastic/plastic, (b) elastic/perfectly plastic, (c) rigid/linear hardening, and (d) rigid/perfectly plastic (see Fig. 6.3).

It is evident in Fig. 6.3(a–d) that yielding occurs when deformation crosses the elastic limit so as to begin plastic behavior. The deformation behaviors (shown in Fig. 6.3) are based on some assumptions which include: (a) the material is isotropic,

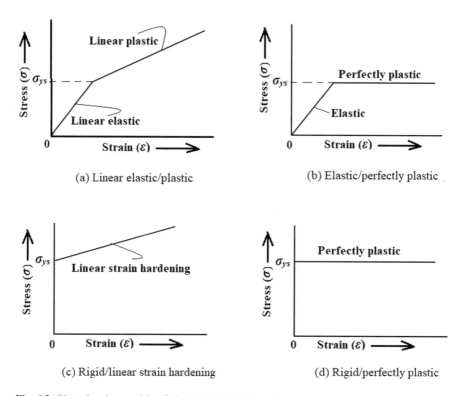

Fig. 6.3 Plots showing models of elastic/plastic deformation

(b) the deformation is independent of rate of loading effects, (c) the yield stress is independent of hydrostatic pressure, and the like. Figure 6.3(a) representing linear elastic/plastic deformation indicates that Hook's law is obeyed up to the proportional/elastic limit, beyond which yielding occurs followed by linear plastic behavior due to strain hardening; this behavior approximates to the deformation in tensile testing of ductile materials (e.g. mild steel). The elastic/perfectly plastic deformation model (Fig. 6.3b) indicates that the yield stress (σ_{ys}) is constant after initial yield i.e. work/strain hardening is ignored; this deformation behavior is exhibited by metals during hot working (HW). Figure 6.3(c) representing rigid/linear strain hardening is indicative of the absence of elastic deformation; which is observed in metal forming processes (e.g. extrusion, wire drawing, etc) involving significantly large strains with up to 50% cold work (CW). The rigid//perfectly plastic deformation behavior (Fig. 6.3d) occurs when the plastic deformation is severe so that the plastic strains are very large compared with the elastic strains – the elastic deformation may be neglected; this deformation behavior is encountered in the engineering analysis of soil and rock stability.

It must be noted that the four deformation models, described in the preceding paragraph, represent simple model; but the deformation behaviors in industrial practice may slightly deviate from the theoretical models. For example, sheet metal forming processes generally involve large deformations together with complex loading sequences. For such situations, Haddag and co-researchers have proposed an advanced, anisotropic elastic–plastic model, formulated within the large strain framework and taking strain-path changes into account, coupled with an isotropic damage model (Haddag, et al. 2009).

6.5   Calcualtions – *Worked Examples*

Example 6.1 Calculating the True Strain in Tensile Loading A 150-mm-long metal rod with a diameter of 2 mm is subject to a tensile force of 1,800 N. As a result, the length increases to 180 mm at failure. Calculate the true strain.

Solution $L_0 = 150$ mm, $L = L_f = 180$ mm, $\varepsilon_{true} = ?$

By using Eq. 6.2,

$$\text{True strain} = \varepsilon_{true} = \ln\left(\frac{L}{L_0}\right) = \ln\left(\frac{180}{150}\right) = \ln 1.2 = 0.182$$

Example 6.2 Calculating the True Stress in Tensile Loading By using the data in Example 6.1, calculate the true stress in the tensile loading.

Solution $L_0 = 150$ mm, $L = L_f = 180$ mm, $F = 1,800$ N, $\sigma_{true} = ?$

$$A_0 = \frac{\pi}{4}d_0^2 = \frac{\pi}{4} \times 2^2 = 3.142 \text{mm}^2$$

By using Eq. 6.4,

$$\text{True stress} = \sigma_{true} = \frac{F*L}{A_0*L_0} = \frac{1800 \times 180}{3.142 \times 150} = 687.46 \, MPa$$

Example 6.3 Using σ_{eng} and ε_{eng} Values to Calculate True Stress and True Strain In a tensile test, the engineering strain is 0.15 and the engineering stress is 720 MPa. Calculate the true stress and the true strain.

Solution $\sigma_{eng} = 720 MPa, \varepsilon_{eng} = 0.15, \sigma_{true} = ?, \varepsilon_{true} = ?$

By using Eqs. 6.5 and 6.6,

$$\text{True stress} = \sigma_{true} = \sigma_{eng}\left(1 + \varepsilon_{eng}\right) = 720(1 + 0.15) = 828 \, MPa$$

True strain $= \varepsilon_{true} = ln\,(1 + \varepsilon_{eng}) = ln\,(1 + 0.15) = 0.14$

Example 6.4 Deriving the Relationship between the Young's-, plastic-, and Tangent Moduli Derive the relationship between Young's modulus, the plastic modulus, and the tangent modulus.

Solution By re-writing Eq. 6.8,

$$d\varepsilon_e = \frac{d\sigma}{E} \qquad \text{(E-6.4a)}$$

By re-writing Eq. 6.9,

$$d\varepsilon_p = \frac{d\sigma}{H} \qquad \text{(E-6.4b)}$$

By re-writing Eq. 6.10,

$$d\varepsilon = \frac{d\sigma}{K} \qquad \text{(E-6.4c)}$$

By combining Eq. E-6.4(a-c) with Eq. 6.11,

$$\frac{d\sigma}{K} = \frac{d\sigma}{H} + \frac{d\sigma}{E}$$

$$\frac{1}{K} = \frac{1}{E} + \frac{1}{H}$$

Example 6.5 Calculating the Plastic Modulus of a Material By using the data in Table 3.1 and Table 6.1, calculate the plastic modulus of annealed copper.

Solution For annealed copper, E = 110,000 MPa (see Table 3.1), and K = 530 MPa (see Table 6.1)

By using the modified form of Eq. 6.12,

$$\frac{1}{H} = \frac{1}{K} - \frac{1}{E} = \frac{1}{530} - \frac{1}{110,000} = 0.001887 - 0.0000091 = 0.001878$$

$$\text{Plastic modulus} = H = \frac{1}{0.001878} = 532.5 \, MPa$$

Example 6.6 Using the Yield Stress value to Calculate the True Strain The strain-hardening stress-strain behavior of annealed 304 stainless steel can be represented by:

$$\sigma_{true} = 1275 \left(\varepsilon_{CW} + 0.002 \right)^{0.45}$$

where σ_{true} is the true stress, and ε_{CW} is the cold-work strain. Calculate the true strain when the yield stress is 750 MPa.

Solution By comparing the given equation with Eq. 6.13, we obtain: $K = 1275$ MPa.

By reference to Fig. 6.1(b), σ_{true} = yield stress = 750 MPa
By substituting σ_{true} = 750 MPa in the given equation,

$$750 = 1275 \left(\varepsilon_{CW} + 0.002 \right)^{0.45}$$

$$\left(\varepsilon_{CW} + 0.002 \right)^{0.45} = 0.588$$

$$\left(\varepsilon_{cw} + 0.002 \right)^{\frac{0.45}{0.45}} = 0.588^{\frac{1}{0.45}}$$

$$\varepsilon_{cw} + 0.002 = 0.588^{2.22} = 0.3076$$

$$\varepsilon_{cw} = 0.3076 - 0.002 = 0.3056$$

By comparing the given equation with Eq. 6.13,

$$\varepsilon_{true} = e_{CW} + 0.002 = 0.3056 + 0.002 = 0.3076$$

True strain = 0.3076

Example 6.7 Calculating the Percent Cold Work by Using the True Strain Value By using the data in *Example 6.6*, calculate the percent cold working that has to be inflicted to induce the required true strain in the steel.

Solution $\varepsilon_{true} = 0.3076, \%CW = ?$

By using Eq. 6.7,

$$\varepsilon_{true} = \ln\left(\frac{A_0}{A}\right)$$

$$0.3076 = \ln\left(\frac{A_0}{A_{CW}}\right)$$

$$\frac{A_0}{A_{CW}} = e^{0.3076} = 1.36$$

By using Eq. 2.6,

$$\%CW = \frac{A_0 - A_{cw}}{A_0} \times 100$$

or $\qquad \%CW = \left[1 - (A_{CW} / A_0)\right] \times 100 = \left[1 - \left(\frac{1}{1.36}\right)\right] \times 100 = 26.4$

The degree of cold work = 26.4%

Example 6.8 Calculating the True Strain by Using n and K values A sample of 2024-T3 aluminum alloy is subjected to a true stress of 480 MPa. Calculate the true strain that will be produced in the material. (Hint: Refer to Table 6.1).

Solution $n = 0.17$ (see Table 6.1), $K = 780$ MPa (see Table 6.1), $\sigma_{true} = 480$ MPa, $\varepsilon_{true} = ?$

By re-writing Eq. 6.13 in logarithmic form,

$$\log \sigma_{true} = \log K + n \log \varepsilon_{true}$$

$$\log 480 = \log 780 + 0.17 \log \varepsilon_{true}$$

$$\log \varepsilon_{rue} = -1.247$$

$\varepsilon_{true} = 10^{-1.247} = 0.0566$

Questions and Problems

6.1. MCQs). Encircle the most appropriate answers for the following statements:

(a) Which modulus refers to the overall increment in strain?
(i) Young's modulus, (ii) plastic modulus, (iii) tangent modulus, (iv) bulk modulus

(b) Which modulus refers to the increment in elastic strain?
(i) Young's modulus, (ii) plastic modulus, (iii) tangent modulus, (iv) bulk modulus

(c) Which deformation behavior is exhibited in wire-drawing metal-forming operation?
(i) Linear elastic/plastic, (ii) elastic/perfectly plastic, (iii) rigid/linear hardening

(d) Which deformation behavior is exhibited by metals during hot working?
(i) Linear elastic/plastic, (ii) elastic/perfectly plastic, (iii) rigid/linear hardening

(e) Which deformation behavior is exhibited by during tensile testing of mild steel?
(i) Linear elastic/plastic, (ii) elastic/perfectly plastic, (iii) rigid/linear hardening

6.2. Draw sketches showing plots of simple models for various elastic/plastic deformations.

6.3. A 170-mm-long metal rod with a diameter of 1.6 mm is subject to a tensile force of 1,800 N. The length increases to 200 mm at failure. Calculate the (a) true strain (b) true stress.

6.4. In a tensile test, the engineering strain is 0.12 and the engineering stress is 680 MPa. Calculate the true stress and the true strain.

6.5. Calculate the percent cold working that has to be inflicted to induce a true strain of 0.28 in a metallic part.

6.6. A sample of 4340 alloy steel is subjected to a true stress of 500 MPa. Calculate the true strain that will be produced in the material. (Hint: Refer to Table 6.1).

6.7. By using the data in Table 3.1 and Table 6.1, calculate the plastic modulus of mild steel.

References

Bertram A, Glüge R (2015) Solid mechanics: theory, modeling, and problems. Springer International Publishing, Cham

Haddag B, Abed-Meraim F, Balan T (2009) Strain localization analysis using a large *deformation* anisotropic *elastic–plastic model* coupled with damage. Int J Plast 25(10):1970–1996

Huda Z (2020) Metallurgy for physicists and engineers. CRC Press, Boca Raton, FL

Huda Z, Bulpett R (2012) Materials science and design for engineers. Trans Tech Publication, Zurich-Durnten

Chapter 7
Elasticity and Viscoelasticity

7.1 Elastic Behavior of Materials

7.1.1 Elasticity and Elastic Constants

In Chaps. 1 and 3, we learnt that *elasticity* refers to the ability of a material to be deformed under load and then return to its original shape and dimensions when the load is removed. In Chap. 1, it was established that according to the Hook's law: *the force needed to extend or compress a spring by some distance (x) is proportional to the distance, up to the elastic limit.* In Chap. 3, we learnt that Young's modulus (E) is an elastic constant of a material that determines its stiffness in tension. The stiffness of materials has great technological importance in designing aerospace structures and steel structures. Another elastic constant is the shear modulus (G); which is the ratio of shear stress to the shear strain (see Chap. 3). Besides E and G, there is another elastic constant – the bulk modulus (B); which is explained in Sect. 7.5. Additionally, there are two other important elastic properties: *Poisson's ratio* (ν) and *resilience* (U_r); these elastic properties are discussed in Sects. 7.2 and 7.3. It should be noted that no material has perfectly elastic behavior; the use of elastic properties such as E and ν should therefore be regarded as a useful approximation.

7.1.2 Anisotropic and Isotropic Materials

The modulus of elasticity (E) values of many materials strongly depend on their crystallographic orientations; such materials are called *anisotropic materials*. When the mechanical behavior of a material is same for all crystallographic orientations, the material is called *isotropic material*. The elastic behavior of *isotropic materials* is expressed in terms of *Poisson's ratio* (ν).

© The Author(s), under exclusive license to Springer Nature Switzerland AG 2022
Z. Huda, *Mechanical Behavior of Materials*, Mechanical Engineering Series,
https://doi.org/10.1007/978-3-030-84927-6_7

7.2 Poisson's Ratio

It is a commonly observed mechanical behavior of metals/materials that an axial elastic elongation, under tension, is accompanied by a corresponding constriction in the lateral direction (see Fig. 7.1). It means that the axial strain is accompanied by a corresponding lateral strain. *Poisson's ratio* (ν) is defined as the ratio of the lateral strain to the axial strain. Mathematically,

$$\nu = -\frac{Lateral\ strain}{Axial\ strain} \tag{7.1}$$

By reference to Fig. 7.1, the axial strain and the lateral strain can be expressed as follows:

$$Axial\ strain = \frac{Axial\ elongation}{Original\ length} = \frac{\Delta L}{L_0} \tag{7.2}$$

$$Lateral\ strain = \frac{Lateral\ contriction}{Original\ diameter} = -\frac{\delta D}{D_0} \tag{7.3}$$

By combining Eqs. 7.1, 7.2, and 7.3, we obtain:

$$Poisson's\ ratio = \nu = -\frac{L_0 \cdot \delta D}{\Delta L \cdot D_0} \tag{7.4}$$

It must be noted here that the reduction in diameter or lateral constriction (δD) is considered for round bars (*e.g.* metal bar/wire drawing operation). In case of metal plate rolling, for example, the increase in length is compensated by a reduction (lateral constriction) in thickness. The significance of Eqs. 7.2, 7.3, and 7.4 is illustrated in Examples 7.1, 7.2, and 7.3. In general, *Poisson's ratio* values for metals lie

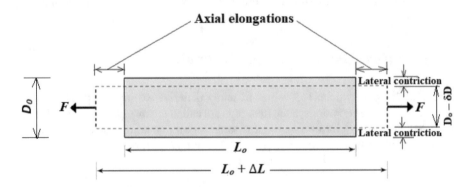

Fig. 7.1 Axial elongation (positive strain) and lateral constriction (negative strain) in a bar

Table 7.1 The elastic constants (v and E) of some materials at ambient temperature

Material	Al_2O_3	PE	Lead	Aluminum	Copper	Titanium	MS	Magnesium
Poisson's ratio (v)	0.22	0.42	0.44	0.345	0.343	0.361	**0.293**	0.29
Young's modulus, GPa	385	0.3	15	70	110	120	212	45

1 GPa = 10^9 N/m^2 (PE = Polyethylene, High Density PE) (MS = Mild Steel)

in the range of 0.28–0.34. The Poisson's ratio and the modulus of elasticity data for various materials are presented in Table 7.1. For isotropic materials, there exists a mathematical relation between Young's modulus (E), the shear modulus (G), and Poisson's ratio (v) according to (Atkin and Fox 2005; Timoshenko 2010):

$$E = 2G(1+v)$$

(7.5)

The significance of Eq. 7.5 is illustrated in Example 7.4.

7.3 Resilience

Resilience of a material refers to the amount of energy absorbed per unit volume when it is deformed elastically and then, upon unloading, to have this energy recovered. Mathematically, *resilience* (U_r) is the area under the stress-strain curve taken to yielding (see Fig. 3.5), or.

$$Ur = \int_{\epsilon_y}^{0} \sigma d\epsilon = \frac{1}{2}\sigma_{ys}\epsilon_{yield}$$

(7.6)

where U_r is the resilience, J/m^3; σ is the stress, Pa; ϵ_{yield} is the elastic strain up to yielding, and σ_{ys} is the yield strength, Pa. By combining Eq. 3.8 and Eq. 7.6, we obtain:

$$U_r = \frac{1}{2}\sigma_{ys}\epsilon_y = \frac{1}{2}\sigma_{ys}\left(\frac{\sigma_{ys}}{E}\right) = \frac{\sigma_{ys}^2}{2E}$$

(7.7)

Equation 7.7 indicates that a resilient material is one having a high yield strength but a low modulus of elasticity (see Example 7.5). Highly resilient alloys find applications in mechanical springs (see also Chap. 1, Sect. 1.2.2.1).

7.4 Generalized Hook's Law – *Hook's Law for Three Dimensions*

In this section, we aim to develop a generalized Hook's law by considering the generalized state of stress at a point, as illustrated in Fig. 7.2. A complete description of stresses consists of normal stresses in three directions (σ_x, σ_y, and σ_z), and shear stresses on three planes (τ_{xy}, τ_{yz}, and τ_{zx}).

Now we analyze the stresses and strains shown in Fig. 7.2. By first considering the normal stresses, the normal stress σ_x causes an axial strain in the x-direction; this strain can be expressed by using Eq. 3.7 as follows:

$$\text{Axial strain caused by } \sigma_x \text{ in } x\text{-direction} = \frac{\sigma_x}{E}$$

$$(7.8)$$

The stress σ_x also causes a (lateral) strain in the y-direction. By considering the strains caused by the stress σ_x in y-direction, Eq. 7.1 can be re-written as (see Fig. 7.2):

$$\text{Lateral strain} = -v\left(\text{axial strain}\right)$$

$$(7.9)$$

By combining Eq. 3.7 and Eq. 7.9 we obtain:

or

$$\text{Lateral strain caused by } \sigma_x \text{ in } y\text{-direction} = -\frac{v\,\sigma_x}{E}$$

$$(7.10)$$

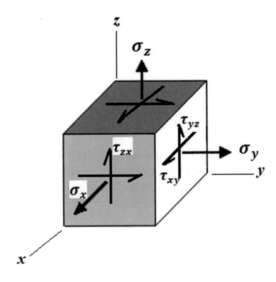

Fig. 7.2 Generalized state of stress at a point showing the six components of stress

Additionally, the stress σ_x also causes a strain in the z-direction; this lateral strain can be expressed, with reference to Eq. 7.10, as follows:

$$\text{Lateral strain caused by } \sigma_x \text{ in } z\text{-direction} = -\frac{v\,\sigma_x}{E}$$

$$(7.11)$$

The total strain caused by the stress σ_x is the sum of the strains in the three directions; this total strain can be obtained by adding the right-hand sides of Eqs. 7.8, 7.10, and 7.11 *i.e.*

$$\text{Total strain caused by the stress } \sigma_x \text{ in } x\text{-,}\,y\text{-, and } z\text{-directions} = \frac{\sigma_x}{E} - \frac{v\,\sigma_x}{E} - \frac{v\,\sigma_x}{E}$$

$$(7.12)$$

The significance of Eq. 7.12 is illustrated in Example 7.6.

Similarly, the normal stresses in the y-direction and z-direction (σ_y and σ_z) each cause strains in all three directions; these strains can be mathematically expressed as:

$$\text{Total strain caused by the stress } \sigma_y \text{ in } x\text{-,}\,y\text{-, and } z\text{-directions} = -\frac{v\,\sigma_y}{E} + \frac{\sigma_y}{E} - \frac{v\,\sigma_y}{E}$$

$$(7.13)$$

$$\text{Total strain caused by the stress } \sigma_z \text{ in } x\text{-,}\,y\text{-, and } z\text{-directions} = -\frac{v\,\sigma_z}{E} - \frac{v\,\sigma_z}{E} + \frac{\sigma_z}{E}$$

$$(7.14)$$

The significance of Eqs. 7.13 and 7.14 is illustrated in Examples 7.7 and 7.8.

Equations 7.12, 7.13, and 7.14 enable us to calculate the total strain, caused by all normal stresses, in each direction, as follows

$$\text{Total strain} \left(\text{caused by } \sigma_x, \sigma_y, \text{ and } \sigma_z \right) \text{ in } x\text{-direction} = \varepsilon_x = \frac{\sigma_x}{E} - \frac{v\,\sigma_y}{E} - \frac{v\,\sigma_z}{E}$$

$$(7.15)$$

or

$$\varepsilon_x = \frac{1}{E}\left[\sigma_x - v\left(\sigma_y + \sigma_z\right)\right]$$

$$(7.16)$$

Similarly, the total strains, caused by σ_x, σ_y, and σ_z, in each of the y- and z-directions are:

$$\varepsilon_y = \frac{1}{E}\left[\sigma_y - v\left(\sigma_x + \sigma_z\right)\right]$$

$$(7.17)$$

$$\varepsilon_z = \frac{1}{E}\left[\sigma_z - v\left(\sigma_x + \sigma_y\right)\right]$$

(7.18)

The significance of Eqs. 7.16, 7.17, and 7.18 is illustrated in Examples 7.9, 7.10, 7.11, 7.12, 7.13, 7.14, 7.15, and 7.16.

The shear strains caused by the shear stresses (shown in Fig. 7.2) can be expressed in terms of the shear modulus, as follows:

$$\gamma_{xy} = \frac{\tau_{xy}}{G}, \quad \gamma_{yz} = \frac{\tau_{yz}}{G}, \quad \gamma_{zx} = \frac{\tau_{zx}}{G}$$

(7.19a–c)

where γ_{xy}, γ_{yz}, and γ_{zx} are the shear strains along the xy, yz, and zx planes, respectively. Eqs. 7.16, 7.17, 7.18, and 7.19a–c are often referred to as the *Generalized Hooke's law* (Ugural and Fenster 1995). The significance of Eq. 7.19a–c is illustrated in Example 7.17.

7.5 Bulk Modulus – *Relationship Between the Elastic Constants*

7.5.1 *Elastic Constants – E, G, and B*

The elastic constant is the ratio of stress to strain. There are three types of elastic constants: (a) Young's modulus, E, (b) shear modulus, G, and (c) bulk modulus, B (Raj and Ramasamy 2012). These elastic constants can simply be expressed as follows:

$$\text{Young's modulus} = E = \frac{Normal\ stress}{Linear\ strain}$$

(7.20)

$$\text{Shear modulus} = G = \frac{Shear\ stress}{Shear\ strain}$$

(7.21)

$$\text{Bulk modulus} = B = \frac{Hydrostatic\ stress}{Volumetric\ strain}$$

(7.22)

The elastic constants E and G have been discussed in detail in Chap. 3. The elastic constant B (bulk modulus) is explained in the following sub-section.

7.5.2   Derivation of Expression for the Bulk Modulus

This section presents the derivation of a mathematical expression for the bulk modulus in terms of the volumetric strain and the hydrostatic stress. The hydrostatic stress and the volumetric strain have a great technological importance in civil/mechanical engineering (Bower 2009).

Volumetric Strain   Consider a stressed rectangular solid, where there are normal strains in x-, y-, and z-directions (see Fig. 7.3).

It is evident in Fig. 7.3 that the elastic deformation of the solid results in the change of dimensions l, w, and h by infinitesimal amounts, dl, dw, and dh, respectively. The corresponding normal strains are:

$$\varepsilon_x = \frac{dl}{l}, \quad \varepsilon_y = \frac{dw}{w}, \quad \varepsilon_z = \frac{dh}{h}$$

$$\text{(7.23a–c)}$$

$$\text{Original volume} = V = l\,w\,h \tag{7.24}$$

The variable V is a function of three independent variables: l, w, and h. By using differential calculus (partial derivatives), the change in volume dV can be determined as follows:

$$dV = \frac{\partial V}{\partial l}\,dl + \frac{\partial V}{\partial w}\,dw + \frac{\partial V}{\partial h}\,dh \tag{7.25}$$

By combining Eqs. 7.24 and 7.25 and using differential calculus principles, we obtain:

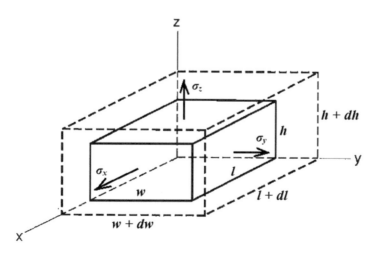

Fig. 7.3   Volume change in a stressed rectangular solid

$$\frac{dV}{V} = \frac{dl}{l} + \frac{dw}{w} + \frac{dh}{h}$$

(7.26)

The term $\dfrac{dV}{V}$ being the ratio of the rate of change of volume to the original volume, is called the *volumetric strain*, ε_v. By combining Eqs. 7.23a–c and 7.26, we get:

$$\frac{dV}{V} = \varepsilon_v = \varepsilon_x + \varepsilon_y + \varepsilon_z$$

(7.27)

Equation 7.27 indicates that the volumetric strain is the sum of the normal strains in the three directions (see Example 7.18).

Hydrostatic Stress The derivation of an expression for the bulk modulus necessitates the definition of hydrostatic stress, as used in Eq. 7.22. The *hydrostatic stress* (σ_h) is the average of the three normal stresses (in x-, y-, and z-directions) *i.e.*

$$\sigma_h = \frac{\sigma_x + \sigma_y + \sigma_z}{3}$$

(7.28)

Equations 7.27 and 7.28 indicate that if the normal stresses and the volume change values are known, the bulk modulus of the material can be calculated by using Eqs. 7.22 (see Example 7.19a–c). Thus Eq. 7.22 can be re-written as

$$B = \frac{Hydrostatic\ stress}{Volumetric\ strain} = \frac{\sigma_h}{\varepsilon_v}$$

(7.29)

The bulk moduli of some materials in engineering are listed in Table 7.2.
By combining Eqs. 7.16, 7.17, and 7.18 and Eq. 7.27, we obtain:

$$\varepsilon_v = \varepsilon_x + \varepsilon_y + \varepsilon_z$$

$$\varepsilon_v = \frac{1}{E}\left[\sigma_x - v\left(\sigma_y + \sigma_z\right)\right] + \frac{1}{E}\left[\sigma_y - v\left(\sigma_x + \sigma_z\right)\right] + \frac{1}{E}\left[\sigma_z - v\left(\sigma_x + \sigma_y\right)\right]$$

$$\varepsilon_v = \frac{1}{E}\left[\sigma_x + \sigma_y + \sigma_z - 2v\left(\sigma_x + \sigma_y + \sigma_z\right)\right]$$

(7.30)

By combining Eqs. 7.28 and 7.30,

$$\varepsilon_v = \frac{3(1-2v)\sigma_h}{E}$$

(7.31)

Table 7.2 Bulk moduli of some materials in engineering at ambient temperature

Material	Aluminum & *Al* alloys	Brass	Bronze	Copper	Cast iron	Carbon Steels	Stainless steel (18–8)	Rubber	Glass	Diamond
Bulk modulus, GPa	68 to 70	108 to 115	112	123	100 to 118	140 to 160	160	1.5 to 2	35 to 55	440

7.5.3   Relationships Between the Elastic Constants and the Poisson's Ratio

The relationship between the bulk modulus (B), the Young's modulus (E), and the Poisson's ratio can be established by combining Eqs. 7.29 and 7.31, as follows:

$$B = \frac{E}{3(1-2v)}$$

(7.32)

The significance of Eq. 7.32 is illustrated in Example 7.20. In order obtain a relationship between E, G, and B, we can re-write Eq. 7.32 as follows:

$$E = 3(1-2v)B$$

(7.33)

By combining Eqs. 7.5 and 7.33, we get:

$$2G(1+v) = 3(1-2v)B$$

(7.34)

In order to eliminate v, we combine Eqs. 7.33 and 7.34 to obtain:

$$E = \frac{9B \cdot G}{G+3B}$$

(7.35)

Equation 7.35 represents the relationship between the three elastic constants – E, G, and B. The significance of Eq. 7.35 is illustrated in Example 7.21.

7.6   Thermal Effects on Elastic Strains

An increase in temperature of a solid causes the atoms to vibrate with a greater amplitude; this thermal effect results in an expansion of the solid (Fig. 7.4). This thermal strain is an elastic strain; which can be expressed as follows:

$$\varepsilon_T = \frac{\delta l}{l} = \alpha(T - T_0) = \alpha(\Delta T)$$

(7.36)

where ε_T is the strain caused by the increase in temperature; T_0 and T are the original and the final temperatures, respectively in °C; and α is the coefficient of thermal expansion, 1/°C (see Example 7.22). The coefficient of thermal expansion is defined as the strain caused by unit rise in temperature.

Thermal strains have a great technological importance. In railway industry, small gaps are allowed between the sections of the rail that form each side of the parallel

Fig. 7.4 Thermal strain in
a metal bar

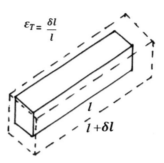

railways. These small gaps provide room for the rails to expand with the rise in temperature due to the friction caused by running of train as well as due to atmospheric temperature rises. It has been experimentally shown that the thermal strain also depends on the rate of heating/cooling (Hassen and Colina 2012)

In isotropic materials, uniform thermal strains are produced in all direction. Thus Eqs. 7.16, 7.17, and 7.18 can be combined with Eq. 7.36 to include thermal effects on strains, as follows:

$$\varepsilon_{x(T)} = \frac{1}{E}\left[\sigma_x - v\left(\sigma_y + \sigma_z\right)\right] + \alpha\left(\Delta T\right)$$

(7.37a)

$$\varepsilon_{y(T)} = \frac{1}{E}\left[\sigma_y - v\left(\sigma_x + \sigma_z\right)\right] + \alpha\left(\Delta T\right)$$

(7.37b)

$$\varepsilon_{z(T)} = \frac{1}{E}\left[\sigma_z - v\left(\sigma_x + \sigma_y\right)\right] + \alpha\left(\Delta T\right)$$

(7.37c)

The significance of Eqs. 7.37a, 7.37b, and 7.37c is illustrated in Example 7.23.

7.7   Viscoelasticity

Viscoelasticity refers to the behavior of materials with both fluid and elastic properties at the same time. It is the viscous-like as well as elastic characteristic of materials. Viscoelasticity is exhibited by almost all polymers; other examples include spaghetti, shag (tobacco), and the like.

The extrusion of plastic parts involves a great deal of viscoelasticity that is manifested by *die swell*. In particular, most elastomers, in the uncured state, exhibit elastic memory – a property of the material to try to return to its originals shape after that shape has been distorted by the application of force. In the processing of an elastomer, the *elastic memory* phenomenon produces *swelling* after a rubber stock has passed through an extruder die. The expansion of hot plastic part when it exits

Fig. 7.5 *Die swell* in
plastic molding (extrusion)

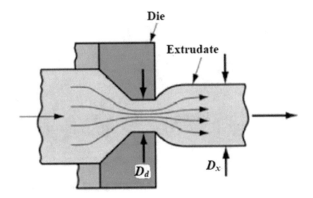

the die opening is called *die swell* (see Fig. 7.5). The die swell problem can be over-
come by using a longer channel in the extrusion machine (Huda 2018).

In circular cross-sectional polymeric parts, *die swell* can be measured by *swell
ratio;* which is defined as the ratio of the diameter of the extrudate to the diameter
of the die orifice (see Fig. 7.5). Mathematically, *swell ratio, r_s,* is given by:

$$r_s = \frac{D_x}{D_d}$$

(7.38)

where D_x is diameter of extrudate; and D_d is diameter of die orifice (see Example 7.24).

7.8 Calculations – *Worked Examples*

Example 7.1 Calculating the Axial Strain by using a Tensile Test Data A 100-
mm long bar, made of an isotropic metal, is acted upon by a tensile stress that results
in an elastic axial elongation of 0.0657 mm. The original diameter of the bar is
10 mm; and the lateral constriction in diameter is 0.002 mm. Calculate the axial
strain in the bar.

Solution $L_0 = 100$ mm; $\Delta L = 0.0657$ mm, Axial strain =?

By using Eq. 7.2

$$\text{Axial strain} = \frac{Axial\,elongation}{Original\,length} = \frac{\Delta L}{L_0} = \frac{0.0657}{100} = 6.57 \times 10^{-4}$$

Example 7.2 Calculating the Lateral Strain by using a Tensile Test Data By
using the data in Example 7.1, calculate the lateral strain in the metal bar.

Solution $\delta D = 0.002$ m; $D = 10$ mm, Lateral strain =?

By using Eq. 7.3,

$$\text{Lateral strain} = -\frac{\text{Lateral contriction in diameter}}{\text{Original diameter}} = -\frac{\delta D}{D_0} = -\frac{0.002}{10} = -0.0002$$

Example 7.3 Calculating the Poisson's Ratio by using a Tensile Test Data   By using the data in Example 7.1, calculate the Poisson's ratio for the metal.

Solution $L_0 = 100$ mm; $\Delta L = 0.0657$ mm, $\delta D = 0.002$ mm; $D = 10$ mm, $\nu =$?

By using Eq. 7.4,

$$\text{Poisson's ratio} = \nu = \frac{L_0 \, \delta D}{D_0 \, \Delta L} = \frac{100 \times 0.002}{10 \times 0.0657} = 0.3$$

Example 7.4 Computing Poisson's Ratio by using E and G Data   By reference to Table 3.1, calculate Poisson's ratio for lead.

Solution $E = 15$ GPa (see Table 3.1), $G = 6$ GPa (see Table 3.1), $\nu =$?

By using Eq. 7.5,

$$E = 2G(1+\nu)$$

$$\text{Poisson's ratio} = \nu = \frac{E - 2G}{2G} = \frac{15 - (2 \times 6)}{2 \times 6} = 0.25$$

Example 7.5 Selecting a Material for Spring Application   The yield strengths of copper and steel samples were experimentally determined to be 70 MPa and 700 MPa, respectively. By reference to the data in Tables 7.1, compute the resilience of copper and steel; and select one of them for spring application.

For copper: $\sigma_{ys} = 70$ MPa, $E = 110$ GPa $= 110,000$ MPa
By using Eq. 7.7 for copper,

$$U_r = \frac{\sigma_{ys}^2}{2E} = \frac{70^2}{2 \times 110,000} = 0.022\text{MPa} = 0.022 \times 10^6 \, \text{Pa} = 0.022\text{MJ} \, / \, \text{m}^3$$

For steel, $\sigma_{ys} = 700$ MPa, $E = 207$ GPa $= 207,000$ MPa

$$U_r = \frac{\sigma_{ys}^2}{2E} = \frac{700^2}{2 \times 207,000} = 1.18\text{MPa} = 1.18\text{MJ} \, / \, \text{m}^3$$

Fig. 7.6 The stress data acting on a cubic element of metal

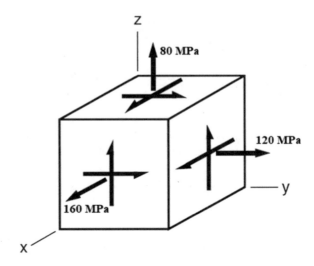

The resilience of steel is much higher than that of copper. Hence, steel must be selected for spring application.

Example 7.6 Calculating the Total Strain Caused by the Stress σ_x in All Three Directions A cubic element of mild steel is acted upon by stresses as shown in Fig. 7.6. Calculate the total strain. Caused by the 160 MPa stress, in all three directions (x-, y-, and z-directions).

Solution σ_x = 160 MPa (see Fig. 7.6), ν = 0.293 (see Table 7.1), $E = 212$ GPa = 212,000 MPa

By using Eq. 7.12,

Total strain caused by the stress σ_x in $x-$, $y-$, and z-directions $= \dfrac{\sigma_x}{E} - \dfrac{\nu\,\sigma_x}{E} - \dfrac{\nu\,\sigma_x}{E}$

Total strain $= \dfrac{\sigma_x}{E}(1-2\nu) = \dfrac{160}{212000}(1-2\times0.293) = 0.000755\times0.414 = 3.12\times10^{-4}$

Total strain caused by the 160 MPa stress in all the three directions $= 3.12\times10^{-4}$

Example 7.7 Calculating the Total Strain Caused by the Stress σ_y in All Three Directions By using the data in Fig. 7.6 for the cubic element of mild steel, calculate the total strain, caused by the 120 MPa stress, in all three directions (x-, y-, and z-directions).

Solution σ_y = 120 MPa (see Fig. 7.6), ν = 0.293 (see Table 7.1), $E = 212$ GPa = 212,000 MPa

By using Eq. 7.13,

Total strain caused by the stress σ_y in $x-, y-,$ and z-directions $= -\dfrac{v\,\sigma_y}{E} + \dfrac{\sigma_y}{E} - \dfrac{v\,\sigma_y}{E}$

Total strain $= \dfrac{\sigma_y}{E}\left(1 - 2v\right) = \dfrac{120}{212000}\left(1 - 2 \times 0.293\right) = 0.000566 \times 0.414 = 2.34 \times 10^{-4}$

Total strain caused by the 120 MPa stress in x , y , and z   directions $= 2.34 \times 10^{-4}$

Example 7.8 Calculating the Total Strain Caused by the Stress σ_z in All Three Directions By using the data in Fig. 7.6, calculate the total strain, caused by the 80 MPa stress, in all three directions (*x*-, *y*-, and *z*-directions).

Solution σ_z = 80 MPa (see Fig. 7.6), v = 0.293 (see Table 7.1), $E = 212$ GPa $= 212,000$ MPa

By using Eq. 7.14,

Total strain caused by the stress σ_z in x-, y-, and z-directions $= -\dfrac{v\,\sigma_z}{E} - \dfrac{v\,\sigma_z}{E} + \dfrac{\sigma_z}{E}$

Total strain $= \dfrac{\sigma_z}{E}\left(1 - 2v\right) = \dfrac{80}{212000}\left(1 - 2 \times 0.293\right) = 0.000377 \times 0.414 = 1.56 \times 10^{-4}$

Total strain, caused by the 80 MPa stress, in$x$, y , and z   directions $= 1.56 \times 10^{-4}$

Example 7.9 Calculating the Total Strain, Caused by all Normal Stresses, in *x*-direction By using the stress data in Fig. 7.6, calculate the total strain, caused by all normal stresses, in *x*-direction.

Solution $\sigma_x = 160$ MPa, $\sigma_y = 120$ MPa, $\sigma_z = 80$ MPa, $\varepsilon_x = $?

By using Equation 7.16,

$$\varepsilon_x = \frac{1}{E}\left[\sigma_x - v\left(\sigma_y + \sigma_z\right)\right] = \frac{1}{212000}\left[160 - 0.293\left(120 + 80\right)\right] = 4.7658 \times 10^{-4}$$

The total strain, caused by all normal stresses, in x direction $= 4.7658 \times 10^{-4}$

It means there is a small increase in the dimension in the x-direction.

Example 7.10 Calculating the Total Strain, Caused by all Normal Stresses, in *y*-direction By using the stress data in Fig. 7.6, calculate the total strain, caused by all normal stresses, in y-direction.

Solution $\sigma_x = 160$ MPa, $\sigma_y = 120$ MPa, $\sigma_z = 80$ MPa, $\varepsilon_y = $?

By using Eq. 7.17,

$$\varepsilon_y = \frac{1}{E}\left[\sigma_y - v\left(\sigma_x + \sigma_z\right)\right] = \frac{1}{212000}\left[120 - 0.293(160 + 80)\right] = 2.33 \times 10^{-4}$$

It means there is a small increase in the dimension in the y-direction.

Example 7.11 Calculating the Total Strain, Caused by all Normal Stresses, in z-direction By using the stress data in Fig. 7.6, calculate the total strain, caused by all normal stresses, in z-direction.

Solution $\sigma_x = 160$ MPa, $\sigma_y = 120$ MPa, $\sigma_z = 80$ MPa, $\varepsilon_z = ?$

By using Eq. 7.18,

$$\varepsilon_z = \frac{1}{E}\left[\sigma_z - v\left(\sigma_x + \sigma_y\right)\right] = \frac{1}{212000}\left[80 - 0.293(160 + 120)\right] = -9.44 \times 10^{-6}$$

It means there is a tiny decrease in the dimension in the z-direction.

Example 7.12 Calculating the Total Strain in Each Direction in a Pressure Vessel Figure 7.7 shows the stresses and the dimensions of a copper pressure vessel with closed ends. The cylindrical vessel is filled with pressurized gas that has generated a longitudinal stress (σ_l) of 180 MPa and a hoop stress of 360 MPa. The stress in the z-direction is negligibly small. Calculate the total strain, caused by all normal stresses, in each direction (x-, y-, and z-directions).

Solution $\sigma_l = \sigma_x = 180$ MPa, $\sigma_h = \sigma_y = 360$ MPa, $\sigma_z = 0$, $\varepsilon_x = ?$, $\varepsilon_y = ?$, $\varepsilon_z = ?$

The cylinder is made of copper so E = 110 GPa = 110,000 MPa, $v = 0.343$ (see Table 7.1)

This problem fits to the case we considered in Sect. 7.4.

By using Eq. 7.16,

$$\varepsilon_x = \frac{1}{E}\left[\sigma_x - v\left(\sigma_y + \sigma_z\right)\right] = \frac{1}{110000}\left[180 - 0.343(360 + 0)\right] = 5.14 \times 10^{-4}$$

By using Eq. 7.17,

$$\varepsilon_y = \frac{1}{E}\left[\sigma_y - v\left(\sigma_x + \sigma_z\right)\right] = \frac{1}{110000}\left[360 - 0.343(180 + 0)\right] = 2.71 \times 10^{-3}$$

By using Eq. 7.18,

$$\varepsilon_z = \frac{1}{E}\left[\sigma_z - v\left(\sigma_x + \sigma_y\right)\right] = \frac{1}{110000}\left[0 - 0.343(180 + 360)\right] = -1.68 \times 10^{-3}$$

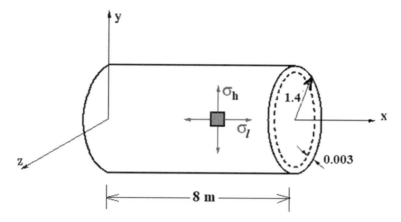

Fig. 7.7 Dimensions of the stressed cylindrical pressure vessel with closed ends

Example 7.13 Determining the Change in Length, Diameter, and Thickness of a Stressed Pressure Vessel By using the data in Example 7.12, determine how much do the length, the diameter, and the wall thickness change. Also specify in each case whether the change is an increase or a decrease.

Solution L_0 = 8 m = 8000 mm, D_0 = 2.8 m = 2800 mm, t = 3 mm, ΔL =?, ΔD =?, Δt =?

By using the modified form of Eq. 7.2,

$$\Delta L = \varepsilon_x L_0 = 5.14 \times 10^{-4} \times 8000 = 4.11 \text{mm}$$

By using the modified form of Equation 7.3,

$$\delta D = \varepsilon_y D_0 = 2.71 \times 10^{-3} \times 2800 = 7.588 \text{mm}$$

Similarly,

$$\delta t = \varepsilon_z t_0 = -1.68 \times 10^{-3} \times 3 = -5.05 \times 10^{-3} \text{mm}$$

To summarize: ΔL = 4.11 mm, δD = 7.588 mm, δt = − 5.05 × 10⁻³mm
    There are small increases in length and diameter, and a very small decrease in the wall thickness.

Example 7.14 Calculating the Stress in a Material that can only Deform in Two Directions In a metal forging operation, a compressive stress of 90 MPa is applied vertically downward (in z-direction) to a sample of mild steel, which is confined in a rigid die. The material cannot deform in the x-direction; it can freely deform in the y-direction. What stress is developed in the x-direction?

Solution σ_z = −90 MPa, σ_y = 0, ε_x = 0, ν = 0.293 (see Table 7.1), $E = 212$ GPa $= 212{,}000$ MPa, $\sigma_x =$?

By applying the generalized Hook's law (using Eq. 7.16)

$$\varepsilon_x = \frac{1}{E}\left[\sigma_x - v\left(\sigma_y + \sigma_z\right)\right]$$

or

$$0 = \frac{1}{212000}\left[\sigma_x - 0.293\left(0 - 90\right)\right]$$

$$0 = \sigma_x + 26.37$$

$$\sigma_x = -26.37 \text{MPa}$$

Example 7.15 Calculating Strain ε_y in a Material that can only Deform in Two Directions By using the data in Example 7.14, calculate the strain in the y-direction.

Solution $\sigma_z = -90$ MPa, $\sigma_y = 0$, $\varepsilon_x = 0$, $\nu = 0.293$, $E = 212$ GPa $= 212{,}000$ MPa, $\varepsilon_y =$?

By using Eq. 7.17,

$$\varepsilon_y = \frac{1}{E}\left[\sigma_y - v\left(\sigma_x + \sigma_z\right)\right] = \frac{1}{212000}\left[0 - 0.293\left(-26.37 - 90\right)\right] = 1.61 \times 10^{-4}$$

The strain in the y-direction $= 1.61 \times 10^{-4}$

Example 7.16 Calculating Strain ε_z in a Material that can only Deform in Two Directions By using the data in Example 7.14, calculate the strain in the z-direction.

Solution $\sigma_z = -90$ MPa, $\sigma_y = 0$, $\varepsilon_x = 0$, $\nu = 0.293$, $E = 212$ GPa $= 212{,}000$ MPa, $\varepsilon_z =$?

By using Eq. 7.18,

$$\varepsilon_z = \frac{1}{E}\left[\sigma_z - v\left(\sigma_x + \sigma_y\right)\right] = \frac{1}{212000}\left[-90 - 0.293\left(-26.37 + 0\right)\right] = -3.88 \times 10^{-4}$$

The strain in the z-direction $= -3.88 \times 10^{-4}$

Example 7.17 Determining the Shear Strains in the Three Planes: xy, yz, and zx The shear stresses of 55 MPa, 50 MPa, and 60 MPa act along xy, yz, and zx planes, respectively, on a cubic solid made of aluminum, as shown in Fig. 7.2. Determine the shear strains in the three planes: xy, yz, and zx.

Solution $\tau_{xy} = 55$ MPa, $\tau_{yz} = 50$ MPa, and $\tau_{zx} = 60$ MPa, $\gamma_{xy} =?$, $\gamma_{yz} =?$, $\gamma_{zx} =?$

By reference to Table 3.1 for aluminum, the shear modulus = $G = 25$ GPa = 25,000 MPa.
By using Eq. (7.19a–c),

$$\gamma_{xy} = \frac{\tau_{xy}}{G} \frac{55}{25000} = 0.0022$$

$$\gamma_{yz} = \frac{\tau_{yz}}{G} = \frac{50}{25000} = 0.0020$$

$$\gamma_{zx} = \frac{\tau_{zx}}{G} = \frac{60}{25000} = 0.0024$$

The shear strains along the planes: *xy*, *yz*, and *zx* are 0.0022, 0.002, and 0.0024, respectively.

Example 7.18 Calculating the Volumetric Strain in a Rectangular Solid   A rectangular solid is subjected to direct stresses that cause volume change as shown in Fig. 7.8. Calculate the volumetric strain in the solid.

Solution $l = 10$ cm, $w = 7$ cm, $h = 3$ cm, $l + dl = 10.8$ cm, $w + dw = 7.5$ cm, $h + dh = 3.2$ cm, $\varepsilon_v =?$

$dl = 10.8–10 = 0.8$ cm, $dw = 0.5$ cm, $dh = 0.2$ cm
By using Eq. 7.23a-c,

$$\varepsilon_x = \frac{dl}{l} = \frac{0.8\,cm}{10\,cm} = 0.08$$

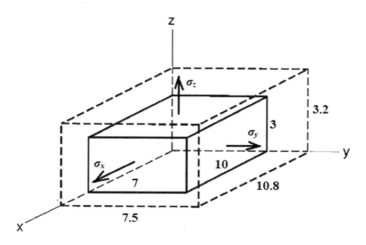

Fig. 7.8 Volume change in a stresses solid (all dimensions in cm)

$$\varepsilon_y = \frac{dw}{w} = \frac{0.5\,cm}{7\,cm} = 0.071$$

$$\varepsilon_z = \frac{dh}{h} = \frac{0.2}{3} = 0.066$$

By using Eq. 7.27,

$$\varepsilon_v = \varepsilon_x + \varepsilon_y + \varepsilon_z = 0.08 + 0.071 + 0.066 = 0.218$$

The volumetric strain = 0.218

Example 7.19 Calculating the Bulk Modulus when the Direct Stresses are Given Refer to the data in Example 7.18. The normal stresses acting on the solid (shown in Fig. 7.8) are: σ_x = 400 MPa, σ_y = 450 MPa, σ_z = 380 MPa. (a) Calculate the bulk modulus for the material of the stressed solid, and (b) identify the material of the stressed solid.

Solution   (a)   By using Eq. 7.28,

$$\sigma_h = \frac{\sigma_x + \sigma_y + \sigma_z}{3} = \frac{400 + 450 + 380}{3} = 410 Mpa$$

The hydrostatic stress = 410 MPa, and the volumetric strain = 0.218 (see Example 7.18).
By using Eq. 7.22,

$$B = \frac{Hydrostatic\ stress}{Volumetric\ strain} = \frac{410}{0.218} = 1881 Mpa = 1.8 Gpa$$

(b) By reference to the data in Table 7.2, the material of the stressed solid is rubber.

Example 7.20 Verification of the Bulk Modulus Data by using the Relationships between Elastic Constants and Poisson's Ratio Refer to the data for aluminum in Table 7.1. (a) Calculate the bulk modulus of aluminum, (b) Compare your answer with the data in Table 7.2, and comment.

Solution   (a)   In Table 7.1, for aluminum E = 70 GPa, v = 0.345,

By using Eq. 7.32,

$$B = \frac{E}{3(1-2v)} = \frac{70}{3(1-2\times0.345)} = \frac{70}{3(1-0.69)} = 75 GPa$$

(b) The answer B = 75 GPa is in agreement with the data in Table 7.2, B = 70 GPa

Example 7.21 Verification of G value by inputting E and B values Refer to the
E and *B* data in Tables 7.1 and 7.2. (a) Calculate the shear modulus for copper, (b)
Compare your answer in (a) with the data in Table 3.1; and comment.

Solution   (a)   By references to Tables 7.1 and 7.2, for copper:
E = 110 GPa, and *B* = 123 GPa.

By using the modified form of Eq. 7.35,

$$G = \frac{3\,E\,B}{9B - E} = \frac{3 \times 110 \times 123}{9 \times 123 - 110} = \frac{40590}{1107 - 110} = 41\text{GPa}$$

(b) By comparing the answer (B = 41 GPa) with the data for copper in Table 3.1
     (B = 46 GPa), it is found that the answer is in agreement with the data.

Example 7.22 Calculating the Final Length After Increase of Temperature A
30-m long mild steel (MS) rod is heated from 25 °C to 180 °C. What is the final
length if the rod is free to expand? The coefficient of thermal expansion of MS is
11×10^{-6} /°C.

Solution $\Delta T = 180 - 25 = 155$ °C, $\alpha = 11 \times 10^{-6}$ /°C, $L_0 = 30$ m, $L_f = ?$

By using Eq. 7.36,

$$\varepsilon_T = \frac{L_f - L_0}{L_0} = \alpha\left(\Delta T\right)$$

$$\frac{L_f - 30}{30} = 11 \times 10^{-6} \times 155$$

$$L_f = 51150 \times 10^{-6} + 30 = 0.05115 + 30 = 30.05\text{m}$$

**Example 7.23 Calculating the Strain Components in all Three Directions if the
Temperature increases while the given Stresses are present** A cubic element of
mild steel is acted upon by stresses as shown in Fig. 7.6. Calculate the total strain,
caused by all normal stresses, in each direction if the temperature increases by
30 °C while the given stresses are present.

Solution $\sigma_x = 160$ MPa, $\sigma_y = 120$ MPa, $\sigma_z = 80$ MPa, $\nu = 0.293$ (see Table 7.1),
E = 212,000 MPa

$\Delta T = 30$ °C, $\alpha = 11 \times 10^{-6}$ /°C
By using Eqs. 7.37a, 7.37b, and 7.37c),

$$\varepsilon_{x(T)} = \frac{1}{E}\left[\sigma_x - v\left(\sigma_y + \sigma_z\right)\right] + \alpha\left(\Delta T\right) = \frac{1}{212000}\left[160 - 0.293\left(120 + 80\right)\right] + \left(11 \times 10^{-6} \times 30\right)$$

$$\varepsilon_{x(T)} = 4.7658 \times 10^{-4} + 330 \times 10^{-6} = 4.7658 \times 10^{-4} + 3.3 \times 10^{-4} = 8.1 \times 10^{-4}$$

The total strain, caused by all normal stresses, in x - direction = 8.1×10^{-4}

$$\varepsilon_{y(T)} = \frac{1}{E}\left[\sigma_y - v\left(\sigma_x + \sigma_z\right)\right] + \alpha\left(\Delta T\right) = \frac{1}{212000}\left[120 - 0.293\left(160 + 80\right)\right] + 300 \times 10^{-6}$$

$$\varepsilon_{y(T)} = 2.33 \times 10^{-4} + 3.3 \times 10^{-4} = 5.63 \times 10^{-4}$$

The total strain, caused by all normal stresses, in y - direction = 5.63×10^{-4}

$$\varepsilon_{z(T)} = \frac{1}{E}\left[\sigma_z - v\left(\sigma_x + \sigma_y\right)\right] + \alpha\left(\Delta T\right) = \frac{1}{212000}\left[80 - 0.293\left(160 + 120\right)\right] + 300 \times 10^{-6}$$
$$= -9.44 \times 10^{-6} + 330 \times 10^{-6} = 320.5 \times 10^{-6} = 3.2 \times 10^{-4}$$

The total strain, caused by all normal stresses, in z - direction = 3.2×10^{-4}

Example 7.24 Calculating the Swell Ratio of an Extruder Die for a Viscoelastic Material The diameter of the die orifice of an extrusion machine is 10 mm. In a polymer processing operation, the die swell results in an increase of the mean diameter of the extrudate to 19 mm after exiting the orifice. Compute the swell ratio.

Solution $D_d = 10$ mm, $D_x = 19$ mm, $r_s = ?$

By using Eq. 7.38,

$$\text{Swell ratio} = r_s = \frac{D_x}{D_d} = \frac{19}{10} = 1.9$$

Questions and Problems

7.1. What is the technological importance of elasticity? Define the three elastic constants.
7.2. Distinguish between *anisotropic materials* and *isotropic materials*.
7.3. Define the term *resilience* and hence derive the relationship: Eq. 7.7.
7.4. Prove mathematically that the volumetric strain is the sum of the normal strains in the three directions.

7.5. An 80-mm long bar, made of an isotropic metal, is acted upon by a tensile stress causing an elastic axial elongation of 0.064 mm. The original diameter of the bar is 7 mm; and the change in diameter is 0.0015 mm. Calculate the Poisson's ratio of the bar's material.

7.6. By using the Poisson's ratio and the Young's modulus data for magnesium (in Table 7.1), calculate the shear modulus of magnesium. Compare your answer with the data in Table 3.1.

7.7. The yield strengths of titanium and steel samples were experimentally determined to be 450

MPa and 800 MPa, respectively. By reference to the data in Tables 7.1, compute the resilience of titanium and steel; and select one of them for spring application.

7.8. A cubic element of nickel is acted upon by stresses as shown in Fig. 7.6. Calculate the total strain. Caused by each stress, in all three directions (x-, y-, and z-directions).

7.9. A mild-steel cylindrical vessel is filled with pressurized gas that has generated a longitudinal stress of 200 MPa and a hoop stress of 400 MPa. The stress in the direction normal to these stresses is negligibly small. Calculate the total strain, caused by all normal stresses, in each direction (x-, y-, and z-directions).

7.10. By using the data in 7.9, calculate the change in the length, the diameter, and the wall thickness. Also specify in each case whether the change is an increase or a decrease. The cylinder dimensions are: original length = 7 m, original internal radius = 1.6 m, original thickness = 0.00 m.

7.11. The shear stresses of 70 MPa, 65 MPa, and 80 MPa act along xy, yz, and zx planes, respectively, on a cubic solid made of copper, as shown in Fig. 7.2. Determine the shear strains in the three planes: xy, yz, and zx.

7.12. A rectangular solid is acted upon by normal stresses that cause strains as shown in Fig. 7.3. The dimensional change data are: length changes from 12 to 12.8 cm; width changes from 9 to 9.6 cm; and the height changes from 6 to 6.4 cm. The normal stresses acting on the solid are: σ_x = 500 MPa, σ_y = 400 MPa, σ_z = 360 MPa. Calculate the bulk modulus for the material of the stressed solid.

7.13. A cubic element of copper is acted upon by stresses as shown in Fig. 7.6. Calculate the total strain, caused by all normal stresses, in each direction if the temperature increases by 40 °C while the given stresses are present. The coefficient of thermal expansion of copper is 16×10^{-6} /°C.

7.14. (MCQ). Encircle the most appropriate answers for the following statements

(a) Which physical quantity is involved in calculating the modulus of rigidity?

(i) linear strain, (ii) shear angle, (iii) volumetric strain, (iv) lateral strain

(b) Which physical quantity is involved in calculating the Poisson's ratio?

(i) linear strain, (ii) shear angle, (iii) volumetric strain, (iv) lateral strain

(c) Which physical quantity is involved in calculating the Young's modulus?

(i) linear strain, (ii) shear angle, (iii) volumetric strain, (iv) lateral strain

(d) Which physical quantity is involved in calculating the bulk modulus?

(i) linear strain, (ii) shear angle, (iii) volumetric strain, (iv) lateral strain.

(e) Which mechanical property is the most important in designing springs?

(i) ductility, (ii) plasticity, (iii) resilience, (iv) creep

References

Atkin RJ, Fox N (2005) An introduction to theory of elasticity. Dover Publications, NY

Bower AF (2009) Applied mechanics of solids. CRC Press, Boca Raton, FL

Hassen S, Colina H (2012) Effect of a heating–cooling cycle on elastic strain and Young's modulus of high performance and ordinary concrete. Mater Struct 45:1861–1875

Huda Z (2018) Manufacturing: mathematical models, problems, and solutions. CRC Press, Boca Raton, FL

Raj PP, Ramasamy V (2012) Strength of materials. Pearson Education Inc., India

Timoshenko SP (2010) Theory of Elasticity. McGraw Hill, India

Ugural AC, Fenster SK (1995) Advanced strength and applied elasticity, prentice hall, New Jersey

Chapter 8
Complex/Principal Stresses and Strains

8.1 Complex Stresses

8.1.1 Technological Importance of Complex and Multiple Stresses

Many structures and machine components, during service, are subjected to complex/multiple stresses resulting from the application of forces. These stresses include tension, compression, torsion, bending, shear, a combination of the stresses, and the like (see Fig. 8.1). For example, complex (plane state) stresses act on the surface of a pressure vessel. Another example is the aircraft; here multiple stresses act on various parts of an aircraft during flight. In particular, the wings, fuselage and landing gear of an aircraft are acted upon by tension, compression, shear, bending, and torsion. A good knowledge of these stresses enable an engineer to design the structure/component against failure.

8.1.2 What Is a Complex Stresses Situation?

In a complex stresses situation, the material in the stressed component has direct and shear stresses acting in two or more directions simultaneously (see Fig. 8.2). The design engineer must ensure that the material of the component does not fail as a result of these stresses. In order to accomplish this objective, the engineer must identify the locations where the stresses are the most severe; and hence s/he must determine the maximum stress in the material. For simplicity, it is appropriate to consider and analyze stresses in two dimensions (2D): x and y. This stress analysis leads us to a useful tool – Mohr's Circle of Stresses; thereby enabling us to

© The Author(s), under exclusive license to Springer Nature Switzerland AG 2022   143
Z. Huda, *Mechanical Behavior of Materials*, Mechanical Engineering Series,
https://doi.org/10.1007/978-3-030-84927-6_8

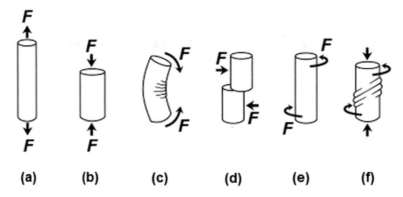

Fig. 8.1 Types of stress; (**a**) tension, (**b**) compression, (**c**) bending, (**d**) shear, (**e**) torsion, and (**f**) combined loading

Fig. 8.2 Complex stresses in plane stress state showing σ_x, σ_y, σ_{xy}, and $\sigma_z = \tau_{yz} = \tau_{zx} = 0$

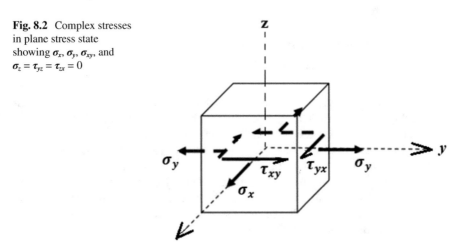

graphically solve complex stress problems (see Sect. 8.5). The 2D complex stresses situation is discussed in the following section.

8.2   The State of Plane Stress – *Axes Transformation*

The state of plane stress exists in many engineering problems; typical examples include: the stresses in metal sheets, the surfaces of thin-walled pressure vessels, the free surfaces of shaft in torsion, beams under transverse load, and the like (Naumenko and Altenbach 2019). Consider a cubic element of material in a solid, and assume that an x-y-z (3D) coordinate system has been chosen for this element. The state of stress is said to be *plane stress* when two faces of the cubic element are free of stress, as shown in Fig. 8.2.

Fig. 8.2 shows three pairs of parallel faces of the element of material acted upon by complex stresses in the state of plane stress. A state of *plane stress* exists if the three components of stress acting on one of the three pairs of parallel faces of the element are all zero *i.e.* $\sigma_z = \tau_{yz} = \tau_{zx} = 0$. In a plane stress problem, the three-dimensional (3D) stress state (Fig. 8.2) can be reduced to two dimensions (2D) (see Fig. 8.3a). Figure 8.3(a) illustrates that three components of stress act on the square element in plane stress state; these stress components are: σ_x, σ_y, and τ_{yx}. A component is said to be under direct loading if either tensile or compressive stresses act on the material. There is a sign convention depending on the type of stress. Tensile normal stresses (Fig. 8.3a) are positive, and compressive normal stresses are negative (see Fig. 8.9). The clockwise shear stresses may be taken as positive, and the counter-clock-wise (CCW) shear stresses are taken as negative.

The state of plane stress described using x-y coordinate (2D) system (Fig. 8.3a) may also be represented on any other coordinate system, such as *1–2* system (Fig. 8.3b); this system is related to the original X-Y coordinate system by an angle of rotation θ. The process of representing a state of stress on a new coordinate system is called *transformation of axes*. The stresses σ_1 and σ_2, shown in Fig. 8.3b, are called the *principal normal stresses* and the axes 1 and 2 are called the *principal axes* (see Sect. 8.4). The principal stresses have a great technological importance in studying the fracture behavior of materials (Yu et al. 2020).

8.2.1   Analyses for Direct and Shear Stresses

In the preceding section, we considered complex stresses on a square element in plane stress state. It is technologically useful to derive equations for direct stresses and shear stresses. Suppose we cut the square element (Fig. 8.3a) in half diagonally

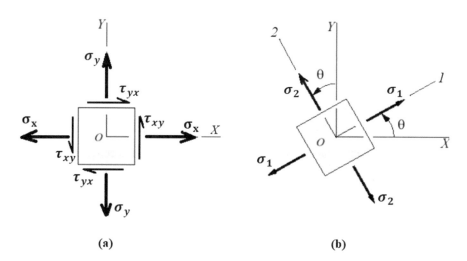

(a)   (b)

Fig. 8.3 A 2D representation of the state of plane stress (**a**); and the principal normal stresses (after axes transformation) (**b**)

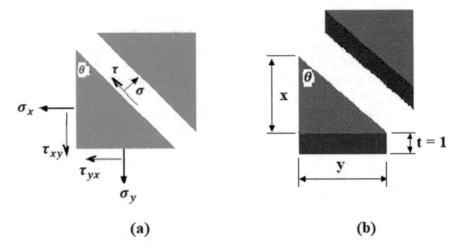

Fig. 8.4 Stress components acting on a portion of the square element (**a**) the 3-D sketch of the element (**b**)

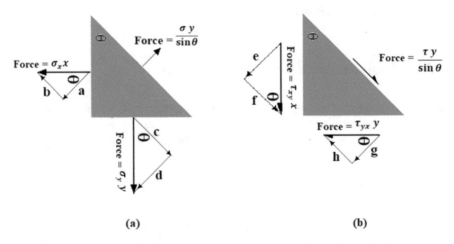

Fig. 8.5 Forces acting on the portion of the square element. (**a**) forces due to the direct stresses, and (**b**) the forces due to the shear stresses

at an angle θ; the resulting portion of the material and the stress components acting on it are shown in Fig. 8.4(a). The stress σ_x acts on the x-plane, and the stress σ_y acts on the y-plane. For convenience, the thickness of the element normal to the diagram is taken as unity (t = 1) (see Fig. 8.4b).

By knowing that force is equal to stress multiplied by the area, the various forces acting on the edges and the sloping plane are shown in Fig. 8.5.

In Fig. 8.5, the forces on the edges have been resolved perpendicular and parallel to the sloping plane. Since the material is in equilibrium, all the forces on the plane must add to zero.

By considering the forces due to direct stresses (in Fig. 8.5a), we obtain,

$$\frac{\sigma\, y}{\sin\theta} - a - d - e - g = 0 \tag{8.1}$$

$$\frac{\sigma\, y}{\sin\theta} = \sigma_x x \cos\theta + \sigma_y y \sin\theta + \tau_{xy} x \sin\theta + \tau_{yx} y \cos\theta \tag{8.2}$$

$$\sigma = \sigma_x \left(\frac{x}{y}\right) \sin\theta \cos\theta + \sigma_y \sin^2\theta + \tau_{xy}\left(\frac{x}{y}\right)\sin^2\theta + \tau_{yx} \sin\theta \cos\theta \tag{8.3}$$

$$\sigma = \sigma_x \left(\frac{\sin\theta \cos\theta}{\tan\theta}\right) + \sigma_y \sin^2\theta + \tau_{xy}\left(\frac{\sin^2\theta}{\tan\theta}\right) + \tau_{yx} \sin\theta \cos\theta \tag{8.4}$$

Since $\tau_{xy} = \tau_{yx}$

$$\sigma = \sigma_x \left(\frac{1 + \cos 2\theta}{2}\right) + \sigma_y \left(\frac{1 - \cos 2\theta}{2}\right) + \tau_{xy}\left(2\sin\theta \cos\theta\right) \tag{8.5}$$

$$\sigma = \frac{\sigma_x + \sigma_y}{2} + \frac{\sigma_x - \sigma_y}{2}\cos 2\theta + \tau_{xy}\sin 2\theta \tag{8.6}$$

By considering the forces due to shear stresses (in Fig. 8.5b), we obtain:

$$\frac{\tau\, y}{\sin\theta} - b + c + f - h = 0 \tag{8.7}$$

By repeating the steps (Eqs. 8.2, 8.3, 8.4, and 8.5), Eq. 8.7 transforms to:

$$\tau = \frac{\sigma_x - \sigma_y}{2}\sin 2\theta - \tau_{xy}\cos 2\theta \tag{8.8}$$

Equation 8.6 and the Eq. 8.8 enable us to calculate the direct stress (σ) and the shear stress (τ) as a function of direction (θ), respectively (see Example 8.1).

8.3   Principal Stresses

In the preceding section, we derived equations for determining the direct stress and the shear stress. A design engineer must ensure that the material of the component does not fail as a result of these stresses. Hence s/he must determine the maximum stresses in the material. The maximum and the minimum values of the direct stress σ and the shear stress τ can be calculated by using the maxima and minima theory (the gradient of a mathematical function is zero at the maximum or minimum point). By taking the derivative $d\sigma/d\theta$ in Eq. 8.6 and equating the result to zero, we obtain:

$$\frac{d\sigma}{d\theta} = 0 - \left(\sigma_x - \sigma_y\right)\sin 2\theta + 2\tau_{xy}\cos 2\theta = 0 \tag{8.9}$$

$$\left(\sigma_x - \sigma_y\right)\sin 2\theta = 2\tau_{xy}\cos 2\theta \tag{8.10}$$

$$\tan 2\theta = \tan 2\theta_n = \frac{2\tau_{xy}}{\left(\sigma_x - \sigma_y\right)} \tag{8.11}$$

where θ_n is the angle of rotation to the coordinate axes for the principal normal stresses. The solution of Eq. 8.11 yields two answers of θ; these angles are less than 360° and differ by 90° (see Example 8.2). The substitution of θ values in Eq. 8.6 yields the maximum and the minimum values of the direct stress as follows:

$$\sigma_{max} = \sigma_1 = \frac{\sigma_x + \sigma_y}{2} + \frac{\sqrt{\left(\sigma_x - \sigma_y\right)^2 + 4\tau_{xy}^2}}{2} \tag{8.12}$$

$$\sigma_{min} = \sigma_2 = \frac{\sigma_x + \sigma_y}{2} - \frac{\sqrt{\left(\sigma_x - \sigma_y\right)^2 + 4\tau_{xy}^2}}{2} \tag{8.13}$$

The maximum and the minimum values of the direct stress (σ_{max} and σ_{min}), from Eqs. 8.12 and 8.13, are called the *principal normal stresses* (see Examples 8.3 and 8.4).

By repeating the mathematical operations for Eq. 8.8, we obtain:

$$\tan 2\theta_s = \frac{\sigma_y - \sigma_x}{2\tau_{xy}} \tag{8.14}$$

$$\tau_{max} = \frac{\sqrt{\left(\sigma_x - \sigma_y\right)^2 + 4\tau_{xy}^2}}{2} = \frac{\sigma_1 - \sigma_2}{2} \tag{8.15}$$

$$\tau_{min} = -\frac{\sqrt{\left(\sigma_x - \sigma_y\right)^2 + 4\tau_{xy}^2}}{2} = -\frac{\sigma_1 - \sigma_2}{2} \tag{8.16}$$

where θ_s is the angle of rotation to the coordinate axes for the principal shear stresses. The maximum and the minimum values of the shear stresses (τ_{max} and τ_{min}), as obtained from Eqs. 8.15 and 8.16, are called the *principal shear stresses* (see

Examples 8.5 and 8.6). The normal stress that accompanies the principal shear stresses can be calculated by:

$$\sigma_{\tau p} = \frac{\sigma_1 + \sigma_2}{2} \tag{8.17}$$

where $\sigma_{\tau p}$ is the normal stress accompanying the principal shear stresses (see Example 8.7).

8.4   Mohr's Circle – *Graphical Representation of Stresses*

Mohr's Circle is a convenient graphical method to determine the principal stresses – this method is widely used in solid mechanics as well as in structural geology (Lahiri and Mamtani 2016). The procedure of developing the Mohr's Circle is given below.

 I. Mark point 'O' at a suitable position, and choose a suitable scale along the horizontal axis.
 II. By using the known values of σ_x and σ_y, represent $\sigma_x = \overline{OP}$ and $\sigma_y = \overline{OQ}$ (see Fig. 8.6).
 III. Mark point 'R' as the mid-point of $\overline{PQ}$.
 IV. In case τ_{xy} is positive, draw τ_{xy} up at P and down at Q. In case τ_{xy} is negative, draw τ_{xy}.
 V. Taking the point R as the center and RS as the radius, draw a circle passing through T.
 VI. Draw the diagonal *ST*. Measure $\overline{OV} = \sigma_1$. Measure $\overline{OU} = \sigma_2$. Measure radius $= \tau_{max}$.
 VII. Measure the angle 2θ as shown in Fig. 8.6.

It is evident from the *Mohr's Circle* (Fig. 8.6) that the principal normal stresses and the principal shear stresses are as follows: $\sigma_1 = \overline{OV}$, $\sigma_2 = \overline{OU}$, $\tau_{max} = +$ radius, $\tau_{min} = -$ radius. The significance of Mohr's Circle is illustrated in Examples 8.8 and 8.9.

8.5   Generalized Plane Stress – *The Presence of σ_z in the Plane Stress*

Consider a three-dimensional view of a state of stress with the presence of the stress σ_z where two components of shear stress are zero *i.e.* $\tau_{yz} = \tau_{zx} = 0$. This state of stress is called the *generalized plane stress*. Fig. 8.7(a) shows the cubic element of material under the *generalized plane stress*. Fig. 8.7(b) illustrates the generalized plane

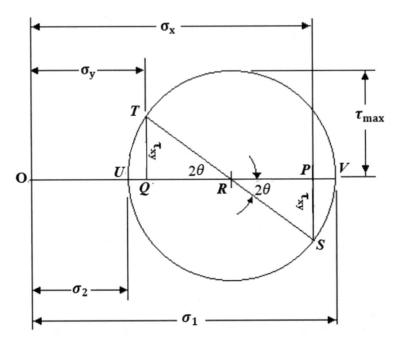

Fig. 8.6 Mohr's Circle of Stress

stress in the x-y plane with the z-direction normal to the plane. Fig. 8.7(c) shows the principal normal stresses (σ_1, σ_2, σ_3) after axes transformation (note that the third principal normal stress turns out to be $\sigma_3 = \sigma_z$).

For generalized plane stress, with $\tau_{yz} = \tau_{zx} = 0$, the principal normal stresses can be calculated either by Mohr's Circle or by using Eqs. 8.12 and 8.13 (see Example 8.10). Additionally, the principal axes can be determined by using Eq. 8.11 (see again Example 8.10).

8.6 Principal Stresses and the Maximum Shear Stress – *3D Consideration*

Consider the principal normal stresses acting on a cubic element of material, as shown in Fig. 8.8(a); here, no shear stress acts on the faces of the cubic element. On a face inclined at any other angle, both normal and shear stresses will act. On faces oriented at 45° with respect to principal axes, it is found that the shear stresses will locally reach a maximum for the three particular directions; these are the maximum shear stress planes (see Fig. 8.8b–d); and the corresponding three shear stresses that result are called the *principal shear stresses*, τ_1, τ_2, and τ_3. The fracture of a uniaxially loaded ductile material occurs owing to the action of the principal shear stresses oriented at 45° with respect to the normal stress axes (see Fig. 1.4a).

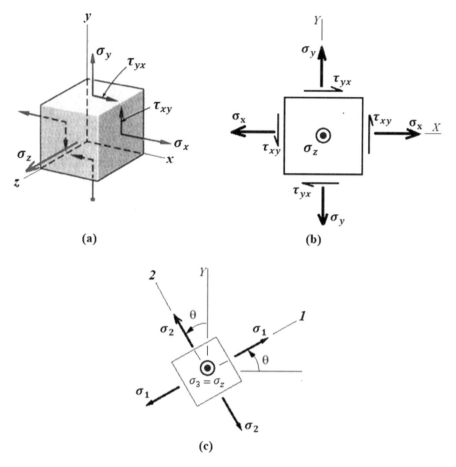

Fig. 8.7 The generalized plane stress acting on a cubic element of material (**a**). the generalized plane stress in the x-y plane with the z-direction normal to the plane (**b**). the principal normal stresses (σ_1, σ_2, σ_3) after axes transformation (**c**)

By reference to Fig. 8.8(b–d), the principal (maximum) shear stresses (τ_1, τ_2, and τ_3) can be expressed in terms of the principal normal stresses, as follows:

$$\tau_1 = \frac{|\sigma_2 - \sigma_3|}{2} \quad \tau_2 = \frac{|\sigma_1 - \sigma_3|}{2} \quad \tau_3 = \frac{|\sigma_1 - \sigma_2|}{2} \tag{8.18a–c}$$

The normal stresses accompanying the shear stresses can be expressed as:

$$\sigma_{\tau 1} = \frac{\sigma_2 + \sigma_3}{2} \quad \sigma_{\tau 2} = \frac{\sigma_1 + \sigma_3}{2} \quad \sigma_{\tau 3} = \frac{\sigma_1 + \sigma_2}{2} \tag{8.19a–c}$$

The significance of Eqs. 8.18a–c and 8.19a–c is illustrated in Examples 8.11 and 8.12.

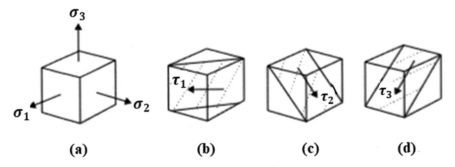

Fig. 8.8 Principal normal stresses (**a**), and the principal shear stresses acting on the maximum shear stress planes (**b–d**)

8.7   Complex Strains – *Principal Strains in 3 Directions*

In Sects. 8.1 and 8.2, we learnt that in a complex stress situation, the material in the stressed component has direct and shear stresses acting in two or more directions simultaneously. We also derived equations for determining the direct and shear stresses; and subsequently derived relationships for calculating the principal stresses (see Sect. 8.4). Since the complex stresses cause complex strains; the principal strains can be calculated by using relationships similar to Eqs. 8.11, 8.12, 8.13, 8.14, and 8.15. Accordingly, the principal normal strains can be computed by:

$$\varepsilon_1 = \frac{\varepsilon_x + \varepsilon_y}{2} + \frac{\sqrt{\left(\varepsilon_x - \varepsilon_y\right)^2 + \gamma_{xy}^2}}{2} \tag{8.20}$$

$$\varepsilon_2 = \frac{\varepsilon_x + \varepsilon_y}{2} - \frac{\sqrt{\left(\varepsilon_x - \varepsilon_y\right)^2 + \gamma_{xy}^2}}{2} \tag{8.21}$$

where ε_1 and ε_2 are the principal normal strains in directions 1 and 2, respectively; γ_{xy} is the shear strain in the xy-plane; and ε_x and ε_y are the linear strains in the x- and y-directions, respectively. For isotropic linear-elastic materials, the third principal normal strain (ε_3) can be computed by:

$$\varepsilon_3 = \varepsilon_z = \frac{v\left(\varepsilon_x + \varepsilon_y\right)}{v - 1} \tag{8.22}$$

The significance of Eqs. 8.20 and 8.22 is illustrated in Example 8.13.

The angle of rotation to the coordinate axes for the principal normal strains can be determined by:

$$\tan 2\theta_n = \frac{\gamma_{xy}}{\varepsilon_x - \varepsilon_y} \tag{8.23}$$

where θ_n the angle of rotation to the coordinate axes for the principal normal strains (see Example 8.14).

Once the principal normal strains have been computed for a 3-dimensional state of strain, the *principal shear strains* can be calculated by (Dowling 2012):

$$\gamma_1 = |\varepsilon_2 - \varepsilon_3|, \gamma_2 = |\varepsilon_1 - \varepsilon_3|, \gamma_3 = |\varepsilon_1 - \varepsilon_2| \tag{8.24a–c}$$

where γ_1, γ_2, γ_3 are the principal shear strains in directions 1, 2, 3, respectively (see Example 8.15). The angle of rotation to the coordinate axes for the strains (γ_1, γ_2, γ_3) can be found by:

$$\tan 2\theta_s = \frac{\varepsilon_y - \varepsilon_x}{\gamma_{xy}} \tag{8.25}$$

where θ_s the angle of rotation to the coordinate axes for the strains (γ_1, γ_2, γ_3) (see Example 8.16). The normal strain accompanying the principal shear strains can be determined by:

$$\varepsilon_{\gamma 3} = \frac{\varepsilon_x + \varepsilon_y}{2} \tag{8.26}$$

where $\varepsilon_{\gamma 3}$ is the normal strain accompanying the principal shear strains (see Example 8.17).

8.8 Calaculations – *Worked Examples*

Example 8.1 Calculating the Stresses (σ and τ) as a Function of Direction (θ) An element of material is acted upon by the direct stresses as shown in Fig. 8.9. Calculate the stresses on a plane inclined 20° to the plane of the tensile direct stress.

Solution According to Fig. 8.9, $\sigma_x = 120$ MPa, $\sigma_y = -80$ MPa, $\tau_{yx} = 0$, $\theta = 20°$, $\sigma = ?, \tau = ?$.

By using Eq. 8.6,

$$\sigma = \frac{\sigma_x + \sigma_y}{2} + \frac{\sigma_x - \sigma_y}{2}\cos 2\theta + \tau_{xy}\sin 2\theta = \frac{120 - 80}{2} + \frac{120 + 80}{2}\cos 40° + 0$$

$$\sigma = 20 + \left(100 \times \cos 40°\right) = 96.6 \, \text{MPa}$$

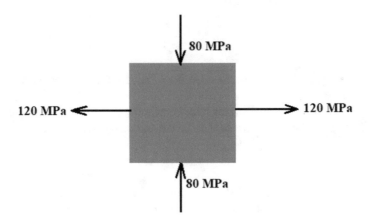

Fig. 8.9 Direct (tensile and compressive) stresses acting on the material

By using Eq. 8.8,

$$\tau = \frac{\sigma_x - \sigma_y}{2}\sin 2\theta - \tau_{xy}\cos 2\theta = \frac{120 + 80}{2}\sin 40° - 0 = 64.2\,\text{MPa}$$

The direct and the shear stresses on the inclined plane are 96.6 MPa and 64.2 MPa, respectively.

Example 8.2 Calculating the Angle to the Coordinate Axes for Principal Normal Stresses A point of interest on the free surface of a machine component is specified. The stresses at the point, with respect to a convenient coordinate system in the plane of the surface, are $\sigma_x = 130$ MPa, $\sigma_y = 210$ MPa, and $\tau_{xy} = 160$ MPa. Determine the angle of rotation to the coordinate axes for the principal normal stresses.

Solution By using Eq. 8.11,

$$\tan 2\theta = \frac{2\tau_{xy}}{\left(\sigma_x - \sigma_y\right)} = \frac{2 \times 160}{130 - 210} = -4$$
$$2\theta = -76° \text{ or } 2\theta = -76 + 180 = 104°$$
$$\theta = -38° \text{ or } \theta = 90 - 38 = 52°$$

The angle of rotation to the coordinate axes for the principal normal stresses is 38° (clock-wise) or 52° (CCW).

Example 8.3 Calculating the Principal Normal Stresses acting on a Machine Component By using the data in Example 8.2, calculate the principal normal stresses acting on the component.

Solution $\sigma_x = 130$ MPa, $\sigma_y = 210$ MPa, and $\tau_{xy} = 160$ MPa, $\sigma_1 = ?$, $\sigma_2 = ?$.

By using Eq. 8.12,

$$\sigma_1 = \frac{\sigma_x + \sigma_y}{2} + \frac{\sqrt{\left(\sigma_x - \sigma_y\right)^2 + 4\tau_{xy}^2}}{2} = \frac{130 + 210}{2} + \frac{\sqrt{\left(130 - 210\right)^2 + \left(4 \times 160^2\right)}}{2}$$

$$\sigma_1 = 170 + \frac{\sqrt{6400 + 102400}}{2} = 170 + 165 = 335 \, \text{MPa}$$

By using Eq. 8.13,

$$\sigma_2 = \frac{\sigma_x + \sigma_y}{2} - \frac{\sqrt{\left(\sigma_x - \sigma_y\right)^2 + 4\tau_{xy}^2}}{2} = \frac{130 + 210}{2} - \frac{\sqrt{\left(130 - 210\right)^2 + \left(4 \times 160^2\right)}}{2}$$

$$\sigma_2 = 170 - 165 = 5 \, \text{MPa}$$

The principal normal stresses are: $\sigma_1 = 335$ MPa, and $\sigma_2 = 5$ MPa.

Example 8.4 Calculating the Shear Stress in a Complex Stress Situation. An element of material is acted upon by direct stresses of 100 MPa tensile and 70 MPa compressive on two mutually perpendicular planes. A clock-wise shear stress acts on the plane with the tensile stress, and an equal and opposite (complimentary) shear stress acts on the other plane. The maximum principal (tensile) stress is 115 MPa. Calculate the shear stress on the planes.

Solution $\sigma_x = 100$ MPa, $\sigma_y = -70$ MPa, $\sigma_{max} = \sigma_1 = 115$ MPa, $\tau_{xy} = \tau_{yx} = ?$.

By using Eq. 8.12,

$$\sigma_{max} = \sigma_1 = \frac{\sigma_x + \sigma_y}{2} + \frac{\sqrt{\left(\sigma_x - \sigma_y\right)^2 + 4\tau_{xy}^2}}{2}$$

$$115 = \frac{100 - 70}{2} + \frac{\sqrt{\left(100 + 70\right)^2 + 4\tau_{xy}^2}}{2}$$

$$\frac{\sqrt{28900 + 4\tau_{xy}^2}}{2} = 100$$

$$28900 + 4\tau_{xy}^2 = 40000$$

$$\tau_{xy} = 52.7 \, \text{MPa}$$

The shear stress on the planes $= \tau_{xy} = \tau_{yx} = 2.7 \, \text{MPa}$

Example 8.5 Calculating the Angle to the Coordinate Axes for Principal Shear Stresses. By using the data in Example 8.2, calculate angle to the coordinate axes for the principal shear stresses.

$\sigma_x = 130$ MPa, $\sigma_y = 210$ MPa, $\tau_{xy} = 160$ MPa, $\theta_s = ?$.

Solution By using Eq. 8.14,

$$\tan 2\theta_s = \frac{\sigma_y - \sigma_x}{2\tau_{xy}} = \frac{210 - 130}{2 \times 160} = 0.25$$

$$2\theta_s = \tan^{-1} 0.25 = 14°$$

The angle to the coordinate axes for the principal shear stresses = 7°.

Example 8.6 Calculating the Principal Shear Stresses acting on a Machine Component. By using the data in Examples 8.2 and 8.3, compute the principal shear stresses acting on the component.

Solution $\sigma_1 = 335$ MPa, $\sigma_2 = 5$ MPa, $\tau_{max} = ?$, $\tau_{min} = ?$.

By using Eq. 8.15,

$$\tau_{max} = \frac{\sigma_1 - \sigma_2}{2} = \frac{335 - 5}{2} = 16\,\text{MPa}$$

By using Eq. 8.16,

$$\tau_{min} = -\frac{\sigma_1 - \sigma_2}{2} = -165\,\text{MPa}$$

The principal shear stresses are 165 MPa and – 165 MPa.

Example 8.7 Calculating the Normal Stress Accompanying the Principal Shear Stresses. By using the data in Example 8.3, calculate the normal stress accompanying the principal shear stresses.

Solution By using Eq. 8.17,

$$\sigma_{\tau p} = \frac{\sigma_1 + \sigma_2}{2} = \frac{335 + 5}{2} = 170\,\text{MPa}$$

The normal stress accompanying the principal shear stresses = 170 MPa.

Example 8.8 Determining the Principal Stresses by using Mohr's Circle A point of interest on the free surface of a machine component is specified. The stresses at the point, with respect to a convenient coordinate system in the plane of

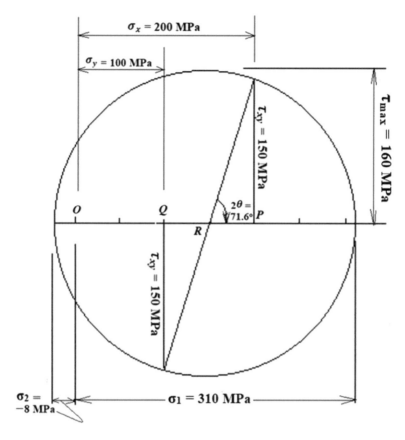

Fig. 8.10   Mohr's Circle for Example 8.8

the surface, are $\sigma_x = 200$ MPa, $\sigma_y = 100$ MPa, and $\tau_{xy} = 150$ MPa. Determine the principal normal stresses and the principal shear stresses by using the Mohr's Circle.

Solution $\sigma_x = 200$ MPa, $\sigma_y = 100$ MPa, and $\tau_{xy} = 150$ MPa, $\sigma_1 = ?$, $\sigma_2 = ?$, $\tau_{max} = ?$, and $\tau_{min} = ?$.

By following the procedure described in Sect. 8.5 (see Steps I – VII), we obtain Fig. 8.10.

By reference to Fig. 8.10, the principal normal stresses are $\sigma_1 = 310$ MPa, and $\sigma_2 = -8$ MPa.

The principal shear stresses are $\tau_{max} = 160$ MPa, and $\tau_{min} = -160$ MPa.

Example 8.9 *Determining* **the Stresses (σ and τ) as a Function of θ by using Mohr's Circle.**   Repeat (solve) Example 8.1 by using Mohr's Circle method.

Solution $\sigma_x = 120$ MPa, $\sigma_y = -80$ MPa, $\sigma = ?$ (for $\theta = 20°$), $\tau = ?$ (for $\theta = 20°$).

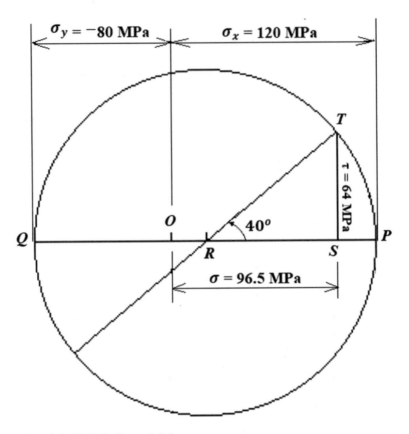

Fig. 8.11 Mohr's Circle for Example 8.9

Mark point 'O' at a suitable position, and choose a suitable scale along the horizontal axis.

Represent $\sigma_x = 120\,\text{MPa} = \overline{OP}$ and $\sigma_y = -80\,\text{MPa} = \overline{OQ}$, as shown in Fig. 8.11.

Draw a circle passing through the points P and Q; and mark the center as R (Fig. 8.11).

By using a protractor taking R as center, draw the required plane at $2\,\theta = 40°$ to the horizontal thereby intersecting the circle at T. Draw the vertical TS. Measure $\overline{OS}$ and $\overline{ST}$.

By reference to Fig. 8.11, the direct stress $= \sigma = 96.5$ MPa, and the shear stress $= \tau = 64$ MPa.

Example 8.10 Determining the Principal Normal stresses and Principal Axes for Generalized Plane Stress A cubic element of material is acted upon by the following generalized plane stress:

$$\sigma_x = 120\,\text{MPa}, \sigma_y = -50\,\text{MPa}, \sigma_z = 60\,\text{MPa}, \tau_{xy} = 70\,\text{MPa}, \tau_{yz} = \tau_{zx} = 0$$

Calculate the (a) principal normal stresses, and (b) the principal axes for the state of stresses.

Solution   (a) By using Eqs. 8.12,

$$\sigma_1 = \frac{\sigma_x + \sigma_y}{2} + \frac{\sqrt{\left(\sigma_x - \sigma_y\right)^2 + 4\tau_{xy}^2}}{2} = \frac{120 - 50}{2} + \frac{\sqrt{\left(120 + 50\right)^2 + 4 \times 70^2}}{2}$$

$$\sigma_1 = \frac{70}{2} + \frac{\sqrt{28900 + 19600}}{2} = 35 + \frac{\sqrt{48500}}{2} = 35 + 110.11 = 145.11\,\text{MPa}$$

By using Eq. 8.13,

$$\sigma_2 = \frac{\sigma_x + \sigma_y}{2} - \frac{\sqrt{\left(\sigma_x - \sigma_y\right)^2 + 4\ \tau_{xy}^2}}{2} = \frac{120 - 50}{2} - \frac{\sqrt{\left(120 + 50\right)^2 + 4 \times 70^2}}{2}$$

$$\sigma_2 = \frac{70}{2} - \frac{\sqrt{28900 + 19600}}{2} = 35 - \frac{\sqrt{48500}}{2} = 35 - 110.11 = -75.1\,\text{MPa}$$

The principal normal stresses are: $\sigma_1 = 145.11$ MPa, $\sigma_2 = -75.1$ MPa, $\sigma_3 = \sigma_z = 60$ MPa

(b) By using Eq. 8.11,

$$\tan 2\theta = \frac{2\tau_{xy}}{\left(\sigma_x - \sigma_y\right)} = \frac{2 \times 70}{\left(120 + 50\right)} = 0.82$$

$$2\theta = 39.4° \text{ or } \theta = 20°$$

The principal axis *1* is 20° CCW of the 120 MPa tensile stress direction, the principal axis *2* is 20° CCW of the 50 MPa compressive stress direction, and the principal axis *3* is in the 60 MPa stress direction (see Fig. 8.7c).

Example 8.11 Determining the *Principal Shear Stresses* **and the** *Maximum Shear Stress*. By using the data in Example 8.10, determine the *principal shear stresses* and the *maximum shear stress* on the material.

Solution $\sigma_1 = 145.11$ MPa, $\sigma_2 = -75.1$ MPa, $\sigma_3 = 60$ MPa, $\tau_1 = ?$, $\tau_2 = ?$, $\tau_3 = ?$, MAX(τ_1, τ_2, τ_3) = ?

By using Eq. (8.18 a–c)

$$\tau_1 = \frac{|\sigma_2 - \sigma_3|}{2} = \frac{|-75.1 - 60|}{2} = \frac{|-135.1|}{2} = \frac{135.1}{2} = 67.55\,\text{MPa}$$

$$\tau_2 = \frac{|\sigma_1 - \sigma_3|}{2} = \frac{|145.11 - 60|}{2} = \frac{|145.11 - 60|}{2} = 42.55\,\text{MPa}$$

$$\tau_3 = \frac{|\sigma_1 - \sigma_2|}{2} = \frac{|145.11 + 75.1|}{2} = 110.1\,\text{MPa}$$

The maximum shear stress, among the three principal shear stresses, is 110.1 MPa.

Example 8.12 Calculating the Normal Stresses Accompanying the Shear Stresses By using the data in Example 8.11, calculate the normal stresses accompanying the shear stresses acting on the material.

Solution $\sigma_1 = 145.11$ MPa, $\sigma_2 = -75.1$ MPa, $\sigma_3 = 60$ MPa, $\sigma_{r1} = ?$, $\sigma_{r2} = ?$, $\sigma_{r3} = ?$.

By using Eq. (8.19a–c),

$$\sigma_{r1} = \frac{\sigma_2 + \sigma_3}{2} = \frac{-75.1 + 60}{2} = -7.55\,\text{MPa}$$

$$\sigma_{r2} = \frac{\sigma_1 + \sigma_3}{2} = \frac{145.11 + 60}{2} = 102.55\,\text{MPa}$$

$$\sigma_{r3} = \frac{\sigma_1 + \sigma_2}{2} = \frac{145.11 - 75.1}{2} = 35\,\text{MPa}$$

The normal stresses accompanying the shear stresses are -7.55 MPa, 102.55 MPa, 102.55 MPa.

Example 8.13 Calculating the Principal Normal Strains in an Isotropic Material There exist the following strains at a point on an unloaded surface of a machine component made of copper: $\varepsilon_x = 0.0045$, $\varepsilon_y = -0.0008$ and $\gamma_{xy} = 0.005$. Assume that the behavior of the material is isotropic linear-elastic. Calculate the principal normal strains.

Solution By using Eqs. 8.20 and 8.21,

$$\varepsilon_1 = \frac{\varepsilon_x + \varepsilon_y}{2} + \frac{\sqrt{\left(\varepsilon_x - \varepsilon_y\right)^2 + \gamma_{xy}^2}}{2} = \frac{0.0045 - 0.0008}{2} + \frac{\sqrt{\left(0.0045 + 0.0008\right)^2 + 0.005^2}}{2}$$
$$\varepsilon_1 = 1.85 \times 10^{-3} + 3.64 \times 10^{-3} = 5.5 \times 10^{3}$$

$$\varepsilon_2 = \frac{\varepsilon_x + \varepsilon_y}{2} - \frac{\sqrt{\left(\varepsilon_x - \varepsilon_y\right)^2 + \gamma_{xy}^2}}{2} = 1.85 \times 10^{-3} - 3.64 \times 10^{-3} = -1.79 \times 10^{-3}$$

Since the material is isotropic linear-elastic ($v = 0.343$), Eq. 8.22 is applicable,

$$\varepsilon_3 = \varepsilon_z = \frac{v\left(\varepsilon_x + \varepsilon_y\right)}{v - 1} = \frac{0.343\left(0.0045 - 0.0008\right)}{0.343 - 1} = \frac{1.27 \times 10^{-3}}{-657 \times 10^{-3}} = -1.93 \times 10^{-3}$$

The principal normal strains are: $\varepsilon_1 = 5.5 \times 10^{-3}$, $\varepsilon_2 = -1.79 \times 10^{-3}$, $\varepsilon_3 = -1.93 \times 10^{-3}$.

Example 8.14 Calculating the Angle of Rotation for the Principal Normal Strains By using the data in Example 8.13, calculate the angle of rotation to the coordinate axes for the.

principal normal strains.

Solution By using Eq. 8.23,

$$\tan 2\theta_n = \frac{\gamma_{xy}}{\varepsilon_x - \varepsilon_y} = \frac{0.005}{0.0045 + 0.0008} = 0.943$$

$$2\theta_n = 43.3°$$

The angle of rotation to the coordinate axes for the principal normal strains = 21.6° (CCW).

Example 8.15 Calculating the Principal Shear Strains in the Three Directions By using the data in Example 8.13, calculate the principal shear strains in the three directions.

Solution $\varepsilon_1 = 5.5 \times 10^{-3}, \varepsilon_2 = -1.79 \times 10^{-3}, \varepsilon_3 = -1.93 \times 10^{-3}, \gamma_1 = ?, \gamma_2 = ?, \gamma_3 = ?.$

By using Eq. 8.24a–c,

$$\gamma_1 = \left|\varepsilon_2 - \varepsilon_3\right| = \left|-17.9 \times 10^{-3} + 1.93 \times 10^{-3}\right| = \left|0.14 \times 10^{-3}\right| = 1.4 \times 10^{-4}$$

$$\gamma_2 = \left|\varepsilon_1 - \varepsilon_3\right| = \left|5.5 \times 10^{-3} + 1.93 \times 10^{-3}\right| = 7.43 \times 10^{-3}$$

$$\gamma_3 = \left|\varepsilon_1 - \varepsilon_2\right| = \left|5.5 \times 10^{-3} 1.79 \times 10^{-3}\right| = 7.29 \times 10^{-3}$$

The principal shear strains are: $\gamma_1 = 1.4 \times 10^{-4}, \gamma_2 = 7.43 \times 10^{-3}, \gamma_3 = 7.29 \times 10^{-3}.$

Example 8.16 Calculating the Angle of Rotation for the Principal Shear Strains By using the data in Example 8.13, calculate the angle of rotation to the coordinate axes for the principal shear strains.

Solution $\varepsilon_x = 0.0045, \varepsilon_y = -0.0008$ and $\gamma_{xy} = 0.005, \theta_s = ?$ (for principal shear strains).

By using Eq. 8.25,

$$\tan 2\theta_s = \frac{\varepsilon_y - \varepsilon_x}{\gamma_{xy}} = \frac{-0.0008 - 0.0045}{0.005} = -1.06$$

$$2\theta_s = \tan^{-1}(-1.06) = -46.7°$$

The angle of rotation to the coordinate axes for the principal shear strains = 23.3° (clock-wise).

Example 8.17 Calculating the Normal Strain Accompanying the Principal Shear Strains By using the data in Example 8.13, calculate the normal strain accompanying the principal shear strains.

Solution By using Eq. 8.26,

$$\varepsilon_{\gamma 3} = \frac{\varepsilon_x + \varepsilon_y}{2} = \frac{0.0045 - 0.0008}{2} = 1.85 \times 10^{-3}$$

The normal strain accompanying the principal shear strains $= 1.85 \times 10^{-3}$

Questions and Problems

8.1. (a) Draw a diagram illustrating the various types of stresses acting on machine components.
(b) Give four examples of machines/components acted upon by complex/multiple stresses.

8.2. (a) Define the terms complex stress, complex strain, and the state of plane stress.
(b) Give a 2D representation of the state of plane stress and analyze the stresses to derive equations for the direct stress and the shear stress.

8.3. (a) Why is the knowledge of principal stress important for a design engineer?
(b) By using differential calculus tools, derive an equation for calculating the angle of rotation to the coordinate axes for the principal normal stresses.
(c) Explain the fracture shown in Fig. 1.4(a) with reference to the principal shear stresses.

8.4. The stresses at a point on the free surface of a component are $\sigma_x = 140$ MPa, $\sigma_y = 210$ MPa, and $\tau_{xy} = 150$ MPa. Calculate: (a) the angle of rotation to the coordinate axes for the principal normal stresses, (b) the principal normal stresses, and (c) the principal shear stresses.

8.5. An element of material is acted upon by a tensile stress of 130 MPa and a compressive stress of 70 MPa on two mutually perpendicular planes. Calculate the stresses on a plane inclined 40° to the plane of the 130 MPa stress.

8.6. Solve *Problem 8.5* by using the Mohr's circle method.

8.7. An element of material is acted upon by direct stresses of 110 MPa tensile and 90 MPa compressive on two mutually perpendicular planes. A clock-wise shear stress acts on the plane with the tensile stress, and an equal and opposite shear stress acts on the other plane. The maximum principal (tensile) stress is 125 MPa. Calculate the shear stress on the planes.

8.8. A cubic element of material is acted upon by the following generalized plane stress:

$\sigma_x = 130$ MPa, $\sigma_y = -70$ MPa, $\sigma_z = 80$ MPa, $\tau_{xy} = 65$ MPa, $\tau_{yz} = \tau_{zx} = 0$

Calculate the (a) principal normal stresses, and (b) the principal axes for the state of stresses.

8.9. There exist the following strains at a point on an unloaded surface of a machine component made of aluminum: $\varepsilon_x = 0.0065$, $\varepsilon_y = -0.0007$ and $\gamma_{xy} = 0.004$. Assume that the behavior of the material is isotropic linear-elastic. Calculate: (a) the principal normal strains, and (b) the angle of rotation to the coordinate axes for the principal normal strains.

8.10. By using the data in *Problem 8.9*, calculate: (a) the principal shear strains in the three directions, and (b) angle of rotation to the coordinate axes for the principal shear strains.

8.11. By using the data in *Problem 8.9*, calculate the normal strain accompanying the shear strains.

References

Dowling NE (2012) Mechanical behavior of materials, 4th edn. Pearson Education, NYC

Lahiri S, Mamtani MA (2016) Scaling the 3-D Mohr circle and quantification of paleostress during fluid pressure fluctuation – application to understand gold mineralization in quartz veins of Gadag (southern India). J Struct Geol 88:63–72

Naumenko K, Altenbach H (2019) Plane stress and plane strain problems. In: Modeling high temperature materials behavior for structural analysis (Eds. Naumenko and Altenbach). Springer Publishing, New York City, pp 137–167

Yu S, Chai L, Yao D, Bao C (2020) Critical ductile fracture criterion based on first principal stress and stress triaxiality. Theor Appl Fract Mech 109

Chapter 9
Plasticity and Superplasticity – *Theory and Applications*

9.1 Plasticity – *Design and Manufacturing Approaches*

Plasticity refers to the ability of a material to be plastically deformed under the action of a stress. Plastic deformation is the deformation beyond the point of yielding during which stresses and strains are no longer proportional. Plastic deformation, being irreversible, is a permanent deformation. It can adversely affect the usefulness of an engineering component by causing large permanent deflections. For designers, plasticity refers to material failure. Plastic deformation frequently occurs in engineering components and may need to be analyzed in design or in determining the cause of a failure. However, for a manufacturing engineer, plasticity refers to formability – ability of a material to be plasticaly deformed. Plasticity enables us to manufacture parts by metal forming (*e.g.* rolling, forging, extrusion, sheet metal forming, and the like) (Huda 2018).

9.2 The Stress-Strain Curve and Plasticity

We learnt in Chap. 3 that a tension test results/data can be used to draw a stress-strain curve; which enables us to determine both elastic and plastic tensile properties. The stress-strain curve in Fig. 9.1 shows that the total strain in any point of a ductile material is the sum of the elastic strain and the plastic strain. From *O* to *A*, the stress-strain relation is linear elastic.

Fig. 9.1 shows the relation between the engineering stress and the engineering strain of annealed mild steel in tension. Until the upper yield point (UYP) is reached at point A, the stress-strain relation *OA* is linear elastic. After that, the yield stress drops to the point B – the lower yield point (LYP); and remains constant up to point C. The stretch BC is called "*plastic yield*" or "*plastic flow*". After point C, the stress

© The Author(s), under exclusive license to Springer Nature Switzerland AG 2022
Z. Huda, *Mechanical Behavior of Materials*, Mechanical Engineering Series,
https://doi.org/10.1007/978-3-030-84927-6_9

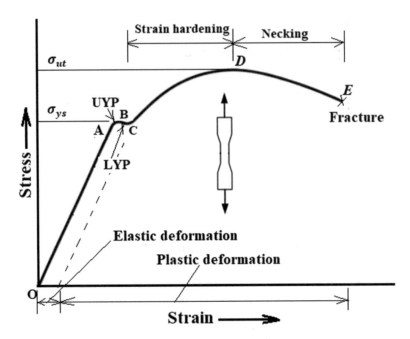

Fig. 9.1 Stress-strain curve for mild steel. (*UYP* = Upper Yield Point, *LYP* = Lower Yield Point)

increases with increasing strain; this phenomenon is called "*work hardening*" or "*strain hardening*". Finally, the maximum engineering stress is reached at point D, after which the stress reduces because of necking of the test sample, until fracture occurs at E (see Sect. 9.3). The stress-strain curve (Fig. 9.1) enables us to determine the following plastic (tensile) properties: the yield strength (σ_{ys}), the ultimate tensile strength (σ_{ut}), the fracture or breaking strength, % elongation, and % reduction in area (see Chap. 3). The yield stress of mild steel is in the order of 200–400 MPa, the ultimate tensile strength is about 400–600 MPa, and the % elongation is in the range of 30–50.

The yielding of steel can be seen on the tensile-test sample by the formation of so-called "*Lüder's lines*", which make an angle of about 45° with the axis of the test sample, indicating that yielding occurs in planes with the greatest shear stress (see Chap. 1, Fig. 1.4a; see Chap. 8, Sect. 8.7). It must be noted that the curve in Fig. 9.1 represents the deformation behavior during loading; however the deformation behavior during unloading differs from that of the behavior for loading due to *Bauschinger effect*; the latter is explained in Sect. 9.4.

9.3   Plastic Instability in Uniaxial Loading

We have learnt in the preceding section that "necking" of a tensile-test sample begins beyond the maximum stress (point D in Fig. 9.1). The necking results in an increase of plastic strain leading to fracture; this phenomenon is called *plastic*

Fig. 9.2 Growth of a neck
indicating plastic
instability

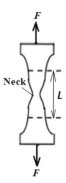

instability. Thus, *plastic instability* refers to the growth of a "neck" in a material upon the application of the maximum stress (Fig. 9.2).

It can be mathematically shown by using Eq. 6.1 and by using differential calculus that:

$$\frac{d\sigma}{\sigma} = -\frac{dA}{A} \tag{9.1}$$

or

$$\frac{d\sigma}{dA} = -\frac{\sigma}{A} \tag{9.2}$$

where σ is the true stress, and A is the instantaneous area of cross-section (see Example 9.1).

By using the constant-volume relationship and by using Eq. 6.2, it can be shown that:

$$-\frac{dA}{A} = \frac{dl}{l} = d\varepsilon \tag{9.3}$$

where l is the gage length at any instant of tensile testing, and ε is the true strain. By combining Eq. 9.3 with Eq. 9.1, we obtain:

$$\frac{d\sigma}{\sigma} = d\varepsilon \tag{9.4}$$

or

$$\frac{d\sigma}{d\varepsilon} = \sigma = \sigma_{ut} \tag{9.5}$$

Equation 9.5 represents the condition of plastic instability *i.e.* the rate of change of true stress with true strain (or rate of strain hardening) is equal to the true stress which corresponds to the ultimate tensile strength (due to necking). The significance of Eqs. 9.1, 9.2, 9.3, 9.4, and 9.5 is illustrated in Examples 9.2, 9.3, and 9.4.

9.4   Bauschinger Effect

In order to study the plastic deformation behavior of materials, it is important to consider stress–strain curves for *monotonic* loading (uniaxial loading). During the tensile loading, the yield strength increases with plastic deformation (see Fig. 9.1); however, during unloading yielding occurs prior to the yield strength. This mechanical behavior of the material (*Bauschinger effect*) was first reported by a German engineer - Johann Bauschinger. The *Bauschinger effect* may be defined as the phenomenon by which plastic deformation of a metal results in an increase in the yield strength in the direction of plastic flow and a decrease in the yield strength in the opposite direction (see Fig. 9.3).

Fig. 9.3 illustrates that if the direction of straining is reversed after yielding has occurred, the stress–strain path differs from the initial monotonic one (pre-strain). This early yielding behavior (lowering of yield stress for reverse loading) is called the *Bauschinger effect*. The main causes of the *Bauschinger effect* are thought to be associated with elastic stress and/or anisotropy in the resistance to dislocation motion. In general, the *Bauschinger effect* has an adverse affect on the mechanical behavior of engineering components; for example, the effective life of automotive suspension springs can be significantly lowered by yielding rather than fracture (Yan 1998).

Fig. 9.3 *Bauschinger effect.* (σ_p = the maximum pre-stress after loading, σ_r = the low yield stress in the reverse loading)

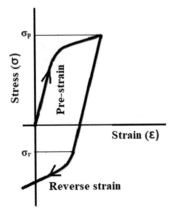

9.5   Bending of Beams – *Plastic Deformation*

9.5.1   Deriving Expressions for the Curvature and the Radius of Curvature

A beam is a one-dimensional structural element. Beam bending is encountered in many application, such as beams in buildings, bridge decks, wind turbine blades, and the like (Krenk and HØgsberg 2013). In bending of a beam, loads act transversely to the longitudinal axis of the beam (Fig. 9.4a). Under the action of loads, shear forces and bending moments are developed; which deform the straight axis of beam into a curve (bent beam) (Fig. 9.4b).

The deflection of the beam is the displacement of the given point from its original position, measured in y direction (see Fig. 9.5). By considering two points p_1 and p_2 on the bent beam, we may write the following mathematical expression:

$$p_1 p_2 = ds = \rho \, d\theta \tag{9.6}$$

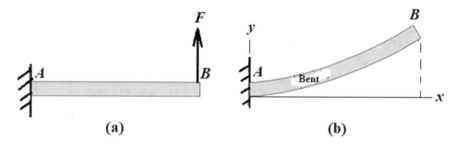

(a)   **(b)**

Fig. 9.4 Bending of beam under the action of a load, F; (**a**) original shape of the beam, (**b**) the beam bent in the y direction

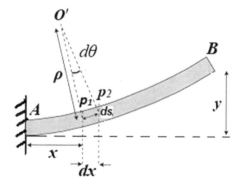

Fig. 9.5 Bending of beam showing the deflection (y), center of curvature (O'), and the radius of curvature (ρ)

where *ds* is the arc length (length of the finite element p_1p_2), mm; ρ is the radius of curvature, mm; and $d\theta$ is the central angle, radians. For small deflections of the beam, $ds \cong dx;$ so Eq. 9.6 takes the form:

$$dx = \rho\, d\theta \qquad (9.7)$$

where *dx* is the distance *ds*, as considered along the *x*-axis (the beam's longitudinal axis). Equation 9.7 can be re-written as,

$$\text{Radius of curvature} = \rho = \frac{dx}{d\theta} \qquad (9.8)$$

The curvature (*k*) is defined as the reciprocal of the radius of curvature *i.e.*

$$\text{Curvature} = k = \frac{1}{\rho} = \frac{d\theta}{dx} \qquad (9.9)$$

The significance of Eqs. 9.6, 9.7, 9.8, and 9.9 is illustrated in Example 9.5.

9.5.2   Symmetrical Bending and the Longitudinal Strain in Simply Supported Beams

Simply supported beams may be acted upon by transverse loading to cause symmetrical bending. Consider a length *AB* of a beam of rectangular cross-section (say) that is subjected to a pure, sagging bending moment, *M*, applied in a vertical plane (see Fig. 9.6). The length AB of the beam will bend into the shape so that the upper surface is concave and the lower convex.

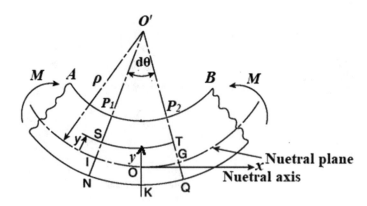

Fig. 9.6 Symmetrical bending in a beam of rectangular cross-section

It is evident in Fig. 9.6 that the element IG lies on the neutral plane and the element of length IO lies along the x-axis (neutral axis). By using Eq. 9.7, the length IG can be expressed as:

$$\overline{IG} = dx = \rho \, d\theta \qquad\qquad\qquad (9.10)$$

After bending,

$$\overline{ST} = L_1 = (\rho - y)d\theta \qquad\qquad (9.11)$$

By combining Eqs. 9.10 and 9.11,

$$L_1 = dx - y\frac{dx}{\rho} \qquad\qquad\qquad (9.12)$$

The surface ST becomes concave *i.e.* the material along ST is compressed; so

$$\text{The change in the length ST} = -\Delta_{ST} = \overline{IG} - \overline{ST} = dx - L_1 \qquad (9.13)$$

$$\Delta_{ST} = L_1 - dx \qquad\qquad\qquad (9.14)$$

By combining Eqs. 9.13 and 9.14, we obtain:

$$\Delta_{ST} = -\frac{y}{\rho}dx \qquad\qquad\qquad (9.15)$$

For small deflections $\overline{ST} = dx$, so the linear strain in the element ST can be calculated by

$$\varepsilon_x = \frac{\Delta_{ST}}{ST} = \frac{-\dfrac{y}{\rho}dx}{dx} = -\frac{y}{\rho} \qquad\qquad (9.16)$$

By combining Eqs. 9.9 and 9.16, we obtain:

$$\varepsilon_x = -\frac{y}{\rho} = -ky \qquad\qquad\qquad (9.17)$$

The negative sign in Eq. 9.17 indicates that the strain is negative when the beam is bent above the neutral axis (y > 0). According to this sign convention, the strain is positive when the beam is bent below the neutral axis (y < 0).

The significance of Eqs. 9.10, 9.11, 9.12, 9.13, 9.14, 9.15, 9.16, and 9.17 is illustrated in Examples 9.6, 9.7, 9.8, and 9.9.

9.6 Application of Plasticity to Sheet Metal Forming

9.6.1 Principal Strain Increments in Uniaxial Loading

The tensile test for flat samples is an example of uniaxial loading, as shown in Fig. 9.7. Since the stress normal to the surface of the sample (direction 3) is negligibly small, the tensile test for a flat sample may be considered in the state of plane stress (see sub-Sect. 9.6.2).

By using the constant volume relationship,

$$l_0 w_0 t_0 = l \cdot w \cdot t = \text{constant} \tag{9.18}$$

It is evident in Fig. 9.7 that:

$$\text{The principal strain increment along the tensile axis} = d\varepsilon_1 = \frac{dl}{l} \tag{9.19a}$$

$$\text{The principal strain increment across the strip } (\text{sample}) = d\varepsilon_2 = \frac{dw}{w} \tag{9.19b}$$

$$\text{The principal strain increment normal to the tensile axis} = d\varepsilon_3 = \frac{dt}{t} \tag{9.19c}$$

The three principal strains can be found by integrating Eqs. 9.19a–c), as follows:

$$\varepsilon_1 = \ln \frac{l}{l_0} \tag{9.20a}$$

$$\varepsilon_2 = \ln \frac{w}{w_0} \tag{9.20b}$$

Fig. 9.7 The gage elements in a tensile test flat-sample showing the principal directions

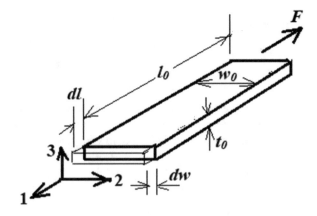

$$\varepsilon_3 = \ln \frac{t}{t_0} \tag{9.20c}$$

The significance of Eqs. 9.18, 9.19a, 9.19b, 9.19c, 9.20a, 9.20b, and 9.20c is illustrated in Examples 9.10 and 9.11.

By using differential calculus for Eq. 9.18, we can write:

or

$$d\left(l_0 w_0 t_0\right) = d\left(l \cdot w \cdot t\right) = 0 \tag{9.21}$$

or

$$d\varepsilon_1 + d\varepsilon_2 + d\varepsilon_3 = 0 \tag{9.22}$$

The proof of Eq. 9.22 is illustrated in Example 9.12.

$$\int d\varepsilon_1 + \int d\varepsilon_2 + \int d\varepsilon_3 = \int 0$$

$$\varepsilon_1 + \varepsilon_2 + \varepsilon_3 = 0 \tag{9.23}$$

It means that the sum of the principal strains in directions 1, 2, and 3 is zero.

9.6.2   Plane Stress Deformation in Sheet Metal Forming

Sheet metal forming processes involve a great deal of plasticity; these processes include: deep drawing (with a flat-headed punch), stretch forming, and general sheet forming (Mellor and Parmar 1978). A common feature of many sheet metal forming process is that the stress, normal to the surface of the sheet, is negligibly small ($\sigma_3 = 0$); thus, the process may be considered to be in the state of *plane stress* (see Chap. 8, Sect. 8.2). The stresses in the plane of the sheet (σ_1 and σ_2) are called the *membrane stresses*. This is why in a typical plane-stress sheet-metal forming process, most work-materials will deform under the membrane stresses (see Fig. 9.8).

The ratio of the stress in direction 2 to the stress in direction 1 is called the stress ratio (α). Similarly, the ratio of strain in direction 2 to the strain in direction 1 is called the strain ratio (β).

$$\alpha = \frac{\sigma_2}{\sigma_1} \tag{9.24}$$

$$\beta = \frac{\varepsilon_2}{\varepsilon_1} = \frac{d\varepsilon_2}{d\varepsilon_1} \tag{9.25}$$

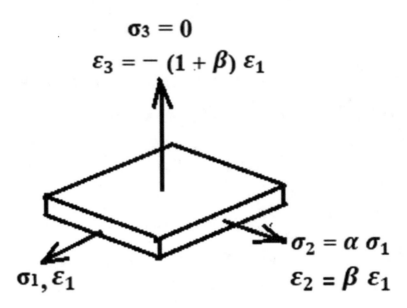

Fig. 9.8 The state of plane stress in sheet metal forming

By combining Eq. 9.23 and Eq. 9.25, we obtain:

$$\varepsilon_3 = -\left(\varepsilon_1 + \varepsilon_2\right) = -\left(\varepsilon_1 + \beta\varepsilon_1\right) = -\left(1 + \beta\right)\varepsilon_1 \qquad (9.26)$$

The significance of Eqs. 9.24, 9.25, and 9.26 is illustrated in Examples 9.13, 9.14, and 9.15.

9.7   Hydrostatic Stress and the Deviatoric Stresses

The hydrostatic stress is the average of the principal normal stresses. By reference to Fig. 8.8(a) (see Chap. 8), the hydrostatic stress (σ_h) is mathematically expressed as:

$$\sigma_h = \frac{\sigma_1 + \sigma_2 + \sigma_3}{3} \qquad (9.27)$$

The deviatoric stress (σ') is the difference between the principal normal stress and the hydrostatic stress (see Fig. 9.9).

It is evident in Fig. 9.9 that the deviatoric stresses (σ_1', σ_2', and σ_3') can be expressed as:

$$\sigma_1' = \sigma_1 - \sigma_h \qquad (9.28a)$$

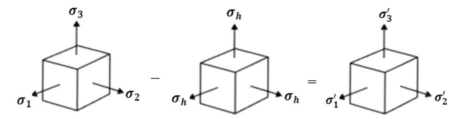

Fig. 9.9 The principal element showing the relationship between the hydrostatic stress (σ_h) and the deviatoric stresses (σ_1', σ_2', and σ_3')

$$\sigma_2' = \sigma_2 - \sigma_h \qquad\qquad (9.28b)$$

$$\sigma_3' = \sigma_3 - \sigma_h \qquad\qquad (9.28c)$$

The significance of Eqs. 9.27, 9.28a, 9.28b, and 9.28c is illustrated in Example 9.16.

In the state of plane stress the deviatoric stresses may be expressed in terms of the stress ratio by:

$$\sigma_1' = \left(\frac{2-\alpha}{3}\right)\sigma_1 \qquad\qquad (9.29a)$$

$$\sigma_2' = \left(\frac{2\alpha-1}{3}\right)\sigma_1 \qquad\qquad (9.29b)$$

$$\sigma_3' = -\left(\frac{1+\alpha}{3}\right)\sigma_1 \qquad\qquad (9.29c)$$

The proof of Eqs. 9.29(a–c) is illustrated in Example 9.17.

9.8   Levy-Mises Flow Rule and Relation Bewteen α and β

Since plasticity is a form of flow, the Levy-Mises flow rule relates the strain rate $\left(\dfrac{d\varepsilon}{dt}\right)$ with stress (σ), as follows:

$$\frac{\dot{\varepsilon}_1}{\sigma_1'} = \frac{\dot{\varepsilon}_2}{\sigma_2'} = \frac{\dot{\varepsilon}_3}{\sigma_3'} \qquad\qquad (9.30)$$

or

$$\frac{\dfrac{d\varepsilon_1}{dt}}{\sigma_1'} = \frac{\dfrac{d\varepsilon_2}{dt}}{\sigma_2'} = \frac{\dfrac{d\varepsilon_3}{dt}}{\sigma_3'} \tag{9.31}$$

or

$$\frac{d\varepsilon_1}{\sigma_1'} = \frac{d\varepsilon_2}{\sigma_2'} = \frac{d\varepsilon_3}{\sigma_3'} \tag{9.32}$$

By combining Eq. 9.29(a–c) and Eq. 9.32,

$$\frac{d\varepsilon_1}{2-\alpha} = \frac{d\varepsilon_2}{2\alpha-1} = \frac{d\varepsilon_3}{-(1+\alpha)} \tag{9.33}$$

or

$$\int\frac{d\varepsilon_1}{2-\alpha} = \int\frac{d\varepsilon_2}{2\alpha-1} = \int\frac{d\varepsilon_3}{-(1+\alpha)} \tag{9.34}$$

or

$$\frac{\varepsilon_1}{2-\alpha} = \frac{\varepsilon_2}{2\alpha-1} = \frac{\varepsilon_3}{-(1+\alpha)} \tag{9.35}$$

By combining Eqs. 9.25, 9.26, and 9.35,
   or

$$\frac{\varepsilon_1}{2-\alpha} = \frac{\varepsilon_2}{2\alpha-1} = \frac{\beta\varepsilon_1}{2\alpha-1} = \frac{\varepsilon_3}{-(1+\alpha)} = \frac{-(1+\beta)\varepsilon_1}{-(1+\alpha)} \tag{9.36}$$

or

$$\alpha = \frac{2\beta+1}{2+\beta} \tag{9.37}$$

where α is the stress ratio and β is the strain ratio (see Example 9.18) .

9.9 Effective Stress and Effective Strain

In soil mechanics, it is important to consider the stress that keeps particles together; this stress is called the *effective stress*. Mathematically, effective stress and effective strain are expressed as:

$$\text{Effective strain} = \bar{\varepsilon} = \sqrt{\frac{4}{3}\left(1 + \beta + \beta^2\right)}\,\varepsilon_1 \tag{9.38}$$

$$\text{Effective stress} = \bar{\sigma} = K\left(a + \bar{\varepsilon}\,\right)^n = \left(\sqrt{1 - \alpha + \alpha^2}\,\right)\sigma_1 \tag{9.39}$$

where K is the strength coefficient, MPa; a and n are constants; ε_1 is the principal strain in direction 1; and σ_1 is the principal normal stress in direction 1 (see Examples 9.19, 9.20, 9.21, and 9.22).

9.10 Superplasticity

We learnt in Sect. 9.2 that the % elongation of mild steel is in the range of 30–50; which indicates a fairly good ductility or plasticity behavior of the material. However, superplastic materials show extra-ordinarily high elongation of over 2000% under tension. *Superplasticity* refers to the plastic deformation of a material involving very high strains at high temperatures ($T > 0.5\,T_m$) before failure; where T_m is the melting temperature in K (Padmanabhan, et al. 2018). Superplasticity, or superplastic forming (SPF) enables us to manufacture complex parts at high temperatures by using a fairly low stress. A typical application of SPF is the manufacture of *Ti-6Al-4 V* aircraft-engine's compressor blades by applying SPF combined with diffusion bonding. For superplastic forming (SPF) of a material, the following microstructural requirements must be fulfilled: (a) a fine grain size of up to 10 µm, (b) equi-axed grains, and (c) multi-phase microstructure (see Fig. 9.10).

Superplasticity is measured as the strain rate sensitivity index (m) which is related to the true stress and the strain rate $\left(\dot{\varepsilon}\right)$ by:

$$\sigma = C\left(\dot{\varepsilon}\right)^m \tag{9.40}$$

where σ is the true stress; C is a constant depending on the temperature; and $\dot{\varepsilon} = \dfrac{d\varepsilon}{dt}$.

By taking logarithm of both sides of Eq. 9.40, we get a linear equation as follows:

$$\log \sigma = m \log \dot{\varepsilon} + \log C \tag{9.41}$$

By plotting a graph of *log σ* vs *log ε̇*, the strain rate sensitivity index (m) can be calculated by:

Fig. 9.10 The microstructure of a typical superplastic forming (SPF) material

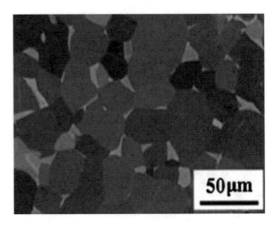

$$m = \text{slope} = \frac{\Delta \log \sigma}{\Delta \log \dot{\varepsilon}} = \frac{\log \sigma_2 - \log \sigma_1}{\log \dot{\varepsilon}_2 - \log \dot{\varepsilon}_1} \tag{9.42}$$

For superplastic forming (SPF), the strain rate sensitivity index (m) should be in the range of 0.4–0.9. The Significance of Eq. 9.42 Is Illustrated in Example 9.23

9.11   Calculations – *Worked Examples*

Example 9.1 Calculating Rate of Change of Stress with Instantaneous Cross-sectional Area A tensile (true) stress of 400 MPa acts on a cross-sectional area of 2 mm² of a test-sample during tensile testing. Calculate the rate of change of stress with respect to instantaneous area of cross-section.

Solution By using Eq. 9.2,

$$\text{Rate of change of stress with area of cross section} = \frac{d\sigma}{dA} = -\frac{\sigma}{A} = -\frac{400}{2} = -200\,\text{N}/\text{mm}^4$$

Example 9.2 Deriving the Relationship: $\dfrac{d\sigma}{\sigma} = -\dfrac{dA}{A}$ By using the definition of true stress, prove Eq. 9.1.

Solution By using the definition of true stress (Eq. 6.1) in the modified form, we get:

$$F = \sigma A$$

Differentiating both sides by using the rule for product of two variables, we obtain:

$$dF = \sigma dA + A d\sigma$$

At the maximum load, F_{max}, there is no change in the load F *i.e.* $dF = 0$ so we get

$$0 = \sigma dA + A d\sigma$$

or

$$A d\sigma = -\sigma dA$$

or

$$\frac{d\sigma}{\sigma} = -\frac{dA}{A}$$

Example 9.3 Deriving the Relationship: $-\dfrac{dA}{A} = \dfrac{dl}{l} = d\varepsilon$ By using the definition of true strain, prove Eq. 9.3.

Solution By using the constant-volume relationship,

$$Al = \text{constant}$$

Differentiating both sides, we get:

$$A\,dl + l\,dA = 0$$

or

$$-\frac{dA}{A} = \frac{dl}{l}$$

By using the definition of true strain (Eq. 6.2), we finally obtain:

$$-\frac{dA}{A} = \frac{dl}{l} = d\varepsilon$$

Example 9.4 Deriving the Condition of Plastic Instability Prove Eq. 9.5.

Solution In Example 9.2, we proved: $\dfrac{d\sigma}{\sigma} = -\dfrac{dA}{A}$.

In Example 9.3, we proved: $-\dfrac{dA}{A} = \dfrac{dl}{l} = d\varepsilon$.

By combining the two equations, we get: $\dfrac{d\sigma}{\sigma} = d\varepsilon$.

or

$$\frac{d\sigma}{d\varepsilon} = \sigma = \sigma_{ut}$$

This is the condition of plastic instability.

Example 9.5 Calculating the Curvature and the Radius of Curvature A beam, under the action of a transverse load, is bent so that the angle subtended by an element of length 2 mm on the beam subtends an angle of 3° with the center of curvature. Calculate the curvature and the radius of curvature.

Solution $d\theta = 3° = \dfrac{\pi}{180} \times 3 = 0.052\,\text{rad}.\ ds = dx = 2\ \text{mm}, \rho = ?, k = ?$

By using Eqs. 9.8 and 9.9,

$$\text{Radius of curvature} = \rho = \frac{dx}{d\theta} = \frac{2}{0.052} = 38.46\,\text{mm}$$

$$\text{Curvature} = k = \frac{1}{\rho} = \frac{1}{38.46} = 0.026\,\text{mm}$$

Example 9.6 Calculating the Length of an Element along the Bent Neutral Plane of Beam In symmetrical bending of a beam, the neutral plane IG is deflected 1.5 mm in y-direction to the length ST (Fig. 9.5). The central angle is 4° and the radius of curvature is 40 mm. Calculate the length ST $\left(\overline{ST} \right)$. The length $\overline{IG}$ is 2.8 mm.

Solution $y = 1.5\ \text{mm},\ d\theta = 4° = \dfrac{\pi}{180} \times 4 = 0.07\,\text{rad}.,\ \rho = 40\ \text{mm},$

By using Eq. 9.11,

$$\overline{ST} = (\rho - y)d\theta = (40 - 1.5) \times 0.07 = 2.695\,\text{mm}$$

Example 9.7 Calculating the Change in Length of Element due to Bending of Beam By using the data in Example 9.6, calculate the change in the length of the element ST due to deflection of the beam.

Solution $dx = \overline{IG} = 2.8\,\text{mm},\ y = 1.5\ \text{mm}, \rho = 40\ \text{mm}, \Delta_{ST} = ?.$

By using Eq. 9.15,

$$\Delta_{ST} = -\frac{y}{\rho}dx = -\frac{1.5 \times 2.8}{40} = -0.105\,\text{mm}$$

Example 9.8 Calculating the Strain in the Bent Element of Beam by Two Methods By using the data in Examples 9.6 and 9.7, calculate the linear strain in the element ST due to bending of beam, as shown in Fig. 9.5. Calculate the strain by two methods.

Solution $\overline{ST} = 2.695\,\text{mm}$ mm (from Example 9.7), $\Delta_{ST} = -0.105$ mm (from Example 9.8), $\varepsilon_x = ?$.

(a) By using the definition of linear strain,

$$\varepsilon_x = \frac{\Delta_{ST}}{ST} = \frac{-0.105}{2.695} = -0.039$$

(b) (b) By using Eq. 9.16,

$$\varepsilon_x = -\frac{y}{\rho} = -\frac{1.5}{40} = -0.0375$$

Example 9.9 Identifying the Mode of Bending in a Beam and Calculating the Curvature A simply supported 5-m-long beam AB with $h = 200$ mm, is acted upon by bending moments M_o, as shown in Fig. 9.11. The beam bends into an arc with strain, $\varepsilon_x = 0.02$. (a) Is the beam bent above or below the neutral axis? (b) Calculate the curvature and the radius of curvature.

Solution   (a) By reference to Eq. 9.12 and the sign convention, the positive strain indicates that the beam is bent below the neutral axis ($y < 0$).
(b) Figure 9.11 shows that the beam surface at the bottom is bent with $y = -100$ mm (the neutral axis passes through the mid-height (mid-thickness).

By using the modified form of Eq. 9.16,

$$\rho = -\frac{y}{\varepsilon_x} = -\frac{-100}{0.02} = 500\,\text{mm} = 5\,\text{m}$$

By using the modified form of Eq. 9.17,

$$k = -\frac{\varepsilon_x}{y} = -\frac{0.02}{-100} = 2 \times 10^{-4}\,\text{mm}$$

Fig. 9.11 A simply-supported 200-mm thick beam under bending moment, M_o

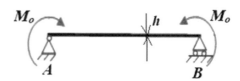

Example 9.10 Calculating the Final Thickness in a Sheet Metal Forming Process A square element 10 mm × 10 mm in an un-deformed metal sheet of 1.2-mm-thickness becomes a rectangle, 9mm × 13mm after forming. Assume that the stress normal to the sheet is zero. Calculate the final thickness of the sheet after the sheet forming process.

Solution $l_0 = 10$ mm, $w_0 = 10$ mm, $t_0 = 1.2$ mm, $l = 13$ mm, $w = 9$ mm, $t = t_f = ?$, $\varepsilon_1 = ?$, $\varepsilon_2 = ?$, $\beta = ?$

By using the modified form of the constant-volume relationship (Eq. 9.18),

$$\text{Final thickness of the sheet} = t_f = t = \frac{l_0\, w_0 t_0}{l\, w} == \frac{10 \times 10 \times 1.2}{13 \times 9} = 1.025\,\text{mm}$$

Example 9.11 Calculating the Principal Strains in a Sheet Metal Forming Process By using the data in Example 9.10, calculate the principal strains in the sheet forming process.

Solution By using Eqs. 9.20(a–c),

$$\text{The principal strain in direction } 1 = \varepsilon_1 = \ln\frac{l}{l_0} = \ln\frac{13}{10} = 0.262$$

$$\text{The principal strain in direction } 2 = \varepsilon_2 = \ln\frac{w}{w_0} = \ln\frac{9}{10} = -0.105$$

$$\text{The principal strain in direction } 3 = \varepsilon_3 = \ln\frac{t}{t_0} = \ln\frac{1.02}{1.2} = -0.1625$$

Example 9.12 Proving that the Sum of the Principal Strain Increments in Directions 1, 2, 3 is Zero Prove Eq. 9.22.

Solution By using Eq. 9.21,

$$d(l \cdot w \cdot t) = 0$$

or

$$l \cdot w\, dt + l \cdot t\, dw + t \cdot w\, dl = 0$$

Dividing both sides by $l \cdot w \cdot t$, we obtain:

$$\frac{dl}{l} + \frac{dw}{w} + \frac{dt}{t} = 0$$

By using Eq. 9.19a, 9.19b, and 9.19c, we get:

$$d\varepsilon_1 + d\varepsilon_2 + d\varepsilon_3 = 0$$

Example 9.13 Determining the Stress Ratio and the Stress in Direction 3 In a plane-stress sheet-metal forming process, the membrane stresses in directions 1 and 2 are 400 MPa and 330 MPa, respectively. Determine the stress ratio and the stress in direction 3.

Solution By using Eq. 9.24,

$$\text{Stress ratio} = \alpha = \frac{\sigma_2}{\sigma_1} = \frac{330\,MPa}{400\,MPa} = 0.825$$

The stress in direction 3 = σ_3 = 0 (since the state is plane stress).

Example 9.14 Calculating the Strain Ratio when the Principal Strains are Given By using the data in Example 9.11, calculate the strain ratio in the sheet forming process.

Solution By using Eq. 9.25,

$$\text{Strain ratio} = \beta = \frac{\varepsilon_2}{\varepsilon_1} = \frac{-0.105}{0.262} = -0.4$$

Example 9.15 Calculating the Strain Ratio when the Principal Strain Increments are Given In a plane-stress sheet-metal forming process, the principal strain increments are 0.014 and 0.009 in the 1 and 2 directions, respectively. Calculate the strain ratio in the sheet forming process.

Solution By using Eq. 9.25,

$$\text{Strain ratio} = \beta = \frac{d\varepsilon_2}{d\varepsilon_1} = \frac{0.009}{0.014} = 0.6428$$

Example 9.16 Calculating the Deviatoric Stresses by using the Principal Normal Stresses The principal normal stresses in directions 1, 2, and 3 are 300 MPa (tensile), 250 MPa (tensile), and 180 MPa (compressive). Calculate the hydrostatic stress and the deviatoric stresses.

Solution By using Eqs. 9.27, 9.28a, 9.28b, and 9.28c,

$$\text{Hydrostatic stress} = \sigma_h = \frac{\sigma_1 + \sigma_2 + \sigma_3}{3} = \frac{300 + 250 - 180}{3} = 123.3\,\text{MPa}$$

The deviatoric stress in direction $1 = \sigma_1' = \sigma_1 - \sigma_h = 300 - 123.3 = 176.67\,\text{MPa}$

The deviatoric stress in direction $2 = \sigma_2' = \sigma_2 - \sigma_h = 250 - 123.3 = 126.7\,\text{MPa}$

The deviatoric stress in direction $3 = \sigma_3' = \sigma_3 - \sigma_h = -180 - 123.3 = -303.3\,\text{MPa}$

Example 9.17 Deriving Expressions for Deviatoric Stresses in Terms of Stress Ratio   Prove Eq. 9.29a, 9.29b, and 9.29c.

Solution $\sigma_1' = \sigma_1 - \sigma_h = \sigma_1 - \left(\dfrac{\sigma_1 + \sigma_2 + \sigma_3}{3}\right) = \dfrac{3\sigma_1 - \sigma_1 - \sigma_2 - 0}{3} = \dfrac{2\sigma_1 - \alpha\sigma_1}{3} = \left(\dfrac{2-\alpha}{3}\right)\sigma_1$

$$\sigma_2' = \sigma_2 - \sigma_h = \sigma_2 - \left(\frac{\sigma_1 + \sigma_2 + \sigma_3}{3}\right) = \frac{3\sigma_2 - \sigma_1 - \sigma_2 - 0}{3} = \frac{2\sigma_2 - \sigma_1}{3} = \frac{2\alpha\sigma_1 - \sigma_1}{3} = \left(\frac{2\alpha - 1}{3}\right)\sigma_1$$

$$\sigma_3' = \sigma_3 - \sigma_h = 0 - \left(\frac{\sigma_1 + \sigma_2 + \sigma_3}{3}\right) = \frac{-\sigma_1 - \sigma_2 - 0}{3} = \frac{-\sigma_1 - \alpha\sigma_1}{3} = -\left(\frac{1+\alpha}{3}\right)\sigma_1$$

Example 9.18 Calculating the Stress Ratio when the Strain Ratio is Known   By using the data in Example 9.14, calculate the stress ratio in the sheet metal forming process.

Solution $\beta = -0.4$, $\alpha =?$

By using Eq. 9.37,

$$\text{Stress ratio} = \alpha = \frac{2\beta + 1}{2 + \beta} = \frac{(2 \times -0.4) + 1}{2 - 0.4} = \frac{-0.8 + 1}{2 - 0.4} = \frac{0.2}{1.6} = 0.125$$

Example 9.19 Calculating the Effective Strain in Sheet Metal Forming Process   A square element 10mm × 10*mm* in an un-deformed metal sheet of 1.2-mm-thickness becomes a rectangle, 9mm × 13mm after forming. Assume that the stress normal to the sheet is zero. The material obeys the following stress-strain law: $\bar{\sigma} = 600(0.008 + \bar{\varepsilon})^{0.22}$ MPa .

Calculate the effective strain for the sheet forming process.

Solution $\varepsilon_1 = \ln\dfrac{13}{10} = 0.262$, $\varepsilon_2 = \ln\dfrac{9}{10} = -0.105$, $\bar{\varepsilon} =?$

$$\beta = \frac{\varepsilon_2}{\varepsilon_1} = \frac{-0.105}{0.262} = -0.4$$

By using Eq. 9.38,

$$\text{Effective strain} = \bar{\varepsilon} = \sqrt{\frac{4}{3}\left(1+\beta+\beta^2\right)}\varepsilon_1 = \sqrt{\frac{4}{3}\left(1-0.4+0.16\right)\times0.262} = 0.5152$$

Example 9.20 Calculating the Effective Stress for the Sheet Forming Process By using the data in Example 9.19, calculate the effective stress for the sheet forming process.

Solution $\bar{\varepsilon} = 0.5152, \bar{\sigma} = 600\left(0.008+\bar{\varepsilon}\,\right)^{0.22}$ MPa.

$$\bar{\sigma} = 600\left(0.008+\bar{\varepsilon}\,\right)^{0.22} = 600\left(0.008+0.5152\,\right)^{0.22} = 520.3\,\text{MPa}$$

Example 9.21 Calculating the Membrane Stresses By using the data in Examples 9.19, 9.20, and 9.20, calculate the final membrane stresses.

Solution $\beta = -0.4$, $\bar{\sigma} = 520.3\,\text{MPa}$, $\sigma_1 = ?$, $\sigma_2 = ?$.

By using Eq. 9.37,

$$\alpha = \frac{2\beta+1}{2+\beta} = \frac{\left(2\times-0.4\right)+1}{2-0.4} = 0.125$$

By using the modified from of Eq. 9.39,

$$\sigma_1 = \frac{\bar{\sigma}}{\sqrt{1-\alpha+\alpha^2}} = \frac{520.3}{\sqrt{1-0.125+0.125^2}} = 551.75\,\text{MPa}$$

By using the modified form of Eq. 9.24,

$$\sigma_2 = \alpha\,\sigma_1 = 0.125\times551.75 = 69\,\text{MPa}$$

Example 9.22 Calculating Hydrostatic Stress and Deviatoric Stresses for Sheet Forming By using the data in Examples 9.19, 9.20, and 9.21, calculate the hydrostatic stress and the deviatoric stresses for the sheet metal forming process.

Solution $\sigma_1 = 551.75$ MPa, $\sigma_2 = 69$ MPa, $\sigma_3 = 0$.

By using Eqs. 9.27, 9.28a, 9.28b, and 9.28c,

$$\text{Hydrostatic stress} = \sigma_h = \frac{\sigma_1 + \sigma_2 + \sigma_3}{3} = \frac{551.75 + 69 + 0}{3} = 207\,\text{MPa}$$

The deviatoric stress in direction $1 = \sigma_1' = \sigma_1 - \sigma_h = 551.75 - 207 = 344.75\,\text{MPa}$

The deviatoric stress in direction $2 = \sigma_2' = \sigma_2 - \sigma_h = 69 - 207 = -138\,\text{MPa}$

The deviatoric stress in direction $3 = \sigma_3' = \sigma_3 - \sigma_h = 0 - 207 = -207\,\text{MPa}$

Example 9.23 Calculating the Strain Rate Sensitivity Index for Checking SPF A series of plasticity experiments were conducted on the MgAZ31 alloy at various stresses and strain rates at a temperature of 350 °C. Two of the data points on the graphical plot are as follows. A true stress of 25 MPa was applied at strain rate of 0.001 s^{-1} while a true stress of 6 MPa was applied at a strain rate of 0.0001 s^{-1}. Calculate the strain rate sensitivity index (m) for the deformation process. Is your computed m value lies in the range for SPF?

Solution $\sigma_1 = 25$ MPa, $\dot{\varepsilon}_1 = 0.0001\text{s}^{-1}$, $\sigma_2 = 6$ MPa, $\dot{\varepsilon}_2 = 0.0001\text{s}^{-1}$, $m = ?$

By using Eq. 9.42,

$$m = \frac{\log \sigma_2 - \log \sigma_1}{\log \dot{\varepsilon}_2 - \log \dot{\varepsilon}_1} = \frac{\log 6 - \log 25}{\log 0.0001 - \log 0.001} = \frac{0.778 - 1.398}{-4 + 3} = 0.62$$

The m = 0.62 value lies in the range for superplastic forming (SPF).

Questions and Problems

9.1. What is the importance of plasticity to a: (a) designer, (b) material processing engineer?

9.2. Draw a stress-strain curve for mild steel, and explain the plasticity region of the curve.

9.3. Why does *Bauschinger effect* have an adverse affect on the mechanical behavior of engineering components (*e.g.* automotive suspension springs)?

9.4. State the condition of plastic, mathematically prove it.

9.5. Define the term "beam". Give three real-life examples of bending of beams.

9.6. Prove that: Linear strain $= -\dfrac{deflection}{radius\ of\ curvature}$

9.7. Derive an expression to calculate the strain rate sensitivity index for SPF.

9.8. In a plane-stress sheet-metal forming process, the principal strain in direction 1 is 0.21 and the strain ratio is 0.58. Calculate the principal strains in directions 2 and 3.

9.9. In symmetrical bending of a beam, the neutral plane IG is deflected 1.2 mm in y-direction to the length ST (Fig. 9.5). The central angle is 3° and the radius of curvature is 35 mm. The length $\overline{IG}$ is 2.6 mm. Calculate the: (a) length $\overline{ST}$, (b) change in the length ST due to deflection of the beam, (c) linear strain in the element ST.

9.10. A simply supported 7-mlong beam AB with h = 300 mm is acted upon by bending moments M_o, as shown in Fig. 9.11. The bending moment cause bending of beam into an arc with strain, $\varepsilon_x = -0.034$. (a) Is the beam bent above or below the neutral axis? (b) Calculate the curvature and the radius of curvature.

9.11. A square element 10mm × 10mm in an un-deformed metal sheet of 1.2-mm-thickness becomes a rectangle, 9mm × 13mm after forming. Assume that the stress normal to the sheet is zero. Calculate the: (a) final thickness of the sheet after the sheet forming process, (b) principal strains in the forming process, (c) strain ratio.

9.12. The principal normal stresses in directions 1, 2, and 3 are 270 MPa (tensile), 200 MPa (compr.), and 170 MPa (tens.). Calculate the hydrostatic stress and the deviatoric stresses.

9.13. A series of plasticity experiments were conducted on the MgAZ31 alloy at various stresses and strain rates at a temperature of 425 °C. Two of the data points on the graphical plot are as follows. A true stress of 10 MPa was applied at strain rate of 0.001 s^{-1} while a true stress of 2 MPa was applied at a strain rate of 0.0001 s^{-1}. Calculate the strain rate sensitivity index (m) for the deformation process. Is your computed m value lies in the range for SPF?

9.14. A square element 12mm × 12mm in an un-deformed metal sheet of 0.9-mm-thickness becomes a rectangle, 8mm × 15mm after forming. Assume that the stress normal to the sheet is zero. The material obeys the following stress-strain law:

$$\bar{\sigma} = 600\left(0.008 + \bar{\varepsilon}\right)^{0.22} \text{ MPa}.$$

Calculate the: (a) effective strain, (b) effective stress, (c) membrane stresses, (d) hydrostatic stress, and (e) deviatoric stresses.

9.15. (MCQs). Encircle the most appropriate answers for the following statements.

(a) Which phenomenon refers to the lowering of yield stress during reverse loading?
(i) plastic instability, (ii) *Bauschinger effect*, (iii) superplasticity, (iv) bending.

(b) Which phenomenon refers to the growth of "neck" of material under uni-axial loading?
(i) bending, (ii) *Bauschinger effect*, (iii) superplasticity, (iv) plastic instability.

(c) Which phenomenon refers to the extra-ordinary high strain at high temperatures?
(i) plastic instability, (ii) *Bauschinger effect*, (iii) superplasticity, (iv) bending.

(d) Which phenomenon involves the curvature and the radius of curvature?
(i) bending, (ii) *Bauschinger effect*, (iii) superplasticity, (iv) plastic instability.

(e) In which condition, the true stress becomes equal to the ultimate tensile strength?
(i) plastic instability, (ii) *Bauschinger effect*, (iii) superplasticity, (iv) bending.

(f) Which stress refers to the average of the principal normal stresses?
(i) membrane stress, (ii) deviatoric stress, (iii) hydrostatic stress, (iv) normal stress.

References

Huda Z (2018) Manufacturing: *mathematical models*. Problems & Solutions, CRC Press, Boca Raton, FL

Krenk S, HØgsberg J (2013) Statics and mechanics of structures. Springer Publishing, Berlin, Germany

Mellor PB, Parmar A (1978) Plasticity analysis of sheet metal forming. In: In: Koistinen and Wang. Mechanics of sheet metal forming. Springer Publishing, Berlin

Padmanabhan KA, Prabu SB, Mulyukov RR, Nazarov A, Imayev RM, Chowdhury SG (2018) Superplasticity. Springer Publishing, Berlin

Yan J (1998) Study of Bauschinger effect on various spring steels, master of applied science thesis. University of Toronto, Graduate Department of Metallurgy & Materials Science

Chapter 10
Torsion in Shafts

10.1 Torsion/Stresses in Shafts

10.1.1 *Torsional Shear Stress in a Shaft*

A shaft is said to be in torsion when it rotates by the application of a torque. Torsion in shafts is found in many industrial applications, especially in drive shafts of vehicles. When a shaft is subjected to a torque, a shear stress (τ) is produced; the latter causes the shaft to rotate about its axis (see Fig. 10.1). The shear stress varies from zero in the axis of rotation to the maximum at the surface of the shaft (Hibbeler 2016).

The shear stress produced in the shaft depends on the applied torque, the distance along the radius of the shaft, and the polar moment of inertia. The shear stress can be calculated by:

$$\tau = \frac{T\,r}{J} \tag{10.1}$$

where τ is the shear stress, Pa; T is the applied torque, N · m; r is the distance along the radius of the shaft, m; and the J is the polar second moment of inertia, m⁴ (see Example 10.1). The *polar moment of inertia of an area* is a measure of a shaft's ability to resist torsion. The polar second moment of inertia (J) for some shaft geometries are given as follows:

For a solid cylindrical shaft,

$$J = \frac{\pi}{32}D^4 \tag{10.2}$$

For a hollow cylindrical shaft,

© The Author(s), under exclusive license to Springer Nature Switzerland AG 2022
Z. Huda, *Mechanical Behavior of Materials*, Mechanical Engineering Series,
https://doi.org/10.1007/978-3-030-84927-6_10

Fig. 10.1 A shaft
subjected to torque
T. (θ = angle of twist,
γ = shear strain)

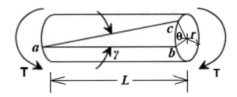

$$J = \frac{\pi}{32}\left(D_o^4 - D_i^4\right) \tag{10.3}$$

For a solid square shaft,

$$J = \frac{a^4}{6} \tag{10.4}$$

10.1.2   Twist and Shear Strain

The shear strain (γ) can be determined by reference to Fig. 10.1, as follows:

$$\gamma = \frac{\widehat{bc}}{L} = \frac{r\theta}{L} \tag{10.5}$$

where θ is the angle of twist, radians; r is the distance along the radius of the shaft; and L is the length of the shaft. Equation 10.5 is valid for both the elastic and plastic deformations of the material. It is important to note that the shear strain and shaft length are inversely proportional to each other? – the longer the shaft, the lower the shear strain (Ghavami 2015).

By the combination of Eq. 3.3, Eqs. 10.1, and 10.5,

$$G = \frac{\tau}{\gamma} = \frac{\dfrac{T\,r}{J}}{\dfrac{r\theta}{L}} = \frac{T\,L}{J\,\theta} \tag{10.6}$$

or

$$\theta = \frac{L\,T}{J\,G} \tag{10.7}$$

The significance of Eqs. 10.2, 10.3, 10.4, 10.5, 10.6, and 10.7 is illustrated in Examples 10.2, 10.3, 10.4, and 10.5 as well as in Examples 10.9, 10.10 & 10.13.

10.1.3 Power and Torque Relationship and Shaft Design

In industrial practice, an electric motor is used to transmit power through a connected solid shaft.

The torque in the shaft is related to the motor-power by:

$$\text{Power} = P = 0.105\, N_{rpm}\, T \tag{10.8}$$

where P is the power of the motor, W; N_{rpm} is the rotational speed of the shaft, rev/min; and T is the torque, N · m (see Example 10.6).

In order to design a shaft, we can derive an expression for the diameter of a solid-cylindrical shaft in terms of the torque and the shear stress by using Eq. 10.1 and Eq. 10.2, as follows:

$$\tau = \frac{T\, r}{J} = \frac{T\dfrac{D}{2}}{\dfrac{\pi}{32}D^4} = \frac{16\,T}{\pi\, D^3} \tag{10.9}$$

or

$$D^3 = \frac{16\,T}{\pi\,\tau} \tag{10.10}$$

or

$$D = 1.72\left(\frac{T}{\tau}\right)^{0.33} \tag{10.11}$$

The significance of Eq. 10.11 is illustrated in Examples 10.7 and 10.8.

10.1.4 Torsional Flexibility and Stiffness

The *torsional flexibility* (f) is defined as the torque per unit twist angle in a shaft. A mathematical expression for the *torsional flexibility* can be obtained by re-writing Eq. 10.7, as follows:

$$\frac{T}{\theta} = \frac{G\,J}{L} \tag{10.12}$$

The terms $\dfrac{T}{\theta}$ is called the *torsional flexibility, f*; thus:

$$\text{Torsional flexibility} = f = \frac{GJ}{L} \tag{10.13}$$

The reciprocal of torsional flexibility is called the *torsional stiffness* (*k*). Mathematically,

$$\text{Torsional stiffness} = k = \frac{1}{f} = \frac{L}{GJ} \tag{10.14}$$

The significance of Eqs. 10.13 and 10.14 is illustrated in Example 10.11.

10.2   Calcualtions – *Worked Examples*

Example 10.1 Calculating the Shear Stress for Torsion in a Solid Shaft A torque of 1200 N · m is acting on a solid cylindrical shaft with diameter 60 mm and length 1.3 m. The shaft is made in steel with modulus of rigidity of 80 GPa. Calculate the maximum shear stress produced in the shaft.

Solution T = 1200 N · m, D = 60 mm = 0.06 m, $r = \dfrac{0.06}{2} = 0.03$ m, L = 1.3 m, G = 80 GPa = 80 × 10⁹ Pa, τ =?

By using Eq. 10.2,

$$J = \frac{\pi}{32} D^4 = \frac{\pi}{32} 0.06^4 = 1.27 \times 10^{-6}\, \text{m}^4$$

Since shear stress is the maximum at the surface of the shaft, we can use Eq. 10.1:

$$\text{The maximum shear stress} = \tau = \frac{Tr}{J} = \frac{1200 \times 0.03}{1.27 \times 10^{-6}} = 28.35 \times 10^6\,\text{Pa} = 28.35\,MPa$$

Example 10.2 Calculating the Angle of Twist for Torsion in a Solid Shaft By using the data in Example 10.1, calculate the angle of twist produced in the shaft.

Solution J = 1.27 × 10⁻⁶ m⁴, T = 1200 N · m, L = 1.3 m, G = 80 GPa = 80 × 10⁹ Pa, θ =?

By using Eq. 10.7,

$$\theta = \frac{LT}{JG} = \frac{1.3 \times 1200}{1.27 \times 10^{-6} \times 80 \times 10^9} = 15.35 \times 10^{-3}\,\text{radians} = \frac{180}{\pi} \times 15.35 \times 10^{-3} = 0.88^\circ.$$

Example 10.3 Calculating the Shear Strain for Torsion in a Solid Shaft by Two Methods By using the data in Examples 10.1 and 10.2, calculate the shear strain in the shaft by two methods.

Solution By using Eq. 10.5,

$$\text{Shear strain} = \gamma = \frac{r\theta}{L} = \frac{0.03 \times 15.35 \times 10^{-3}}{1.3} = 0.000354\,\text{radians} = 0.02^{\circ}$$

By using the modified form of Eq. 3.3,

$$\text{Shear strain} = \gamma = \frac{\tau}{G} = \frac{28.35 \times 10^{6}}{80 \times 10^{9}} = 0.000354$$

Example 10.4 Calculating the Shear Stress for Torsion in a Hollow Shaft A torque of 1200 N · m is acting on a hollow cylindrical shaft with length 1.3 m, outer diameter of 50 mm, and inner diameter of 30 mm. The shaft is made in steel with modulus of rigidity of 80 GPa. Calculate the maximum shear stress produced in the shaft.

Solution T = 1200 N · m, D_o = 50 mm, D_i = 30 mm, $r = \dfrac{50\,mm}{2} = 25\,\text{mm} = 0.025\,\text{m}$
, L = 1.3 m, G = 80 × 10⁹ Pa.

By using Eq. 10.3 (for hollow shaft),

$$J = \frac{\pi}{32}\left(D_o^4 - D_i^4\right) = \frac{\pi}{32}\left(50^4 - 30^4\right) = 534140\,\text{mm}^4 = 0.534 \times 10^{-6}\,\text{m}^4$$

By using Eq. 10.1 (since shear stress is the maximum at the surface of the shaft),

$$\text{The maximum shear stress} = \tau = \frac{Tr}{J} = \frac{1200 \times 0.025}{0.534 \times 10^{-6}} = 56.18 \times 10^{6}\,\text{Pa} = 56.18$$

Example 10.5 Calculating Angle of Twist and Shear Strain for Torsion in Hollow Shaft By using the data in Example 10.4, calculate the angle of twist and the shear strain for torsion in the hollow shaft.

Solution By using Eq. 10.7,

$$\theta = \frac{LT}{JG} = \frac{1.3 \times 1200}{0.534 \times 10^{-6} \times 80 \times 10^{9}} = 36.5 \times 10^{-3}\,\text{radians}$$

By using the modified form of Eq. 3.3,

$$\text{Shear strain} = \gamma = \frac{\tau}{G} = \frac{56.18 \times 10^6}{80 \times 10^9} = 0.7 \times 10^{-3} = 0.0007$$

Example 10.6 Calculating the Torque in a Solid Shaft Connected with an Electric Motor A 12 kW electric motor shall be used to transmit power through a solid connected shaft. The shaft rotates with 1800 rev/min. The machine design requires the maximum allowable shear stress in the shaft to be 75 MPa. Calculate the torque in the shaft.

Solution Power = P = 12 kW = 12,000 W, N_{rpm} = 1800 rev/ min , T = ?.

By using the modified form of Eq. 10.8,

$$\text{Torque} = T = \frac{P}{0.105\, N_{rpm}} = \frac{12000}{0.105 \times 1800} = 63.5\, N \cdot m$$

Example 10.7 Calculating the Diameter of a Solid Shaft in Torsion By using the data in Example 10.6, calculate the diameter of the shaft.

Solution τ = 75 MPa = 75 × 10⁶ Pa, T = 63.5 N · m, D =?,

By using Eq. 10.11,

$$D = 1.72\left(\frac{T}{\tau}\right)^{0.33} = 1.72\left(\frac{63.5}{75 \times 10^6}\right)^{0.33} = 1.72 \times 0.01 \times 0.946 = 0.016\ m = 16mm$$

The solid shaft diameter = 16 mm.

Example 10.8 Calculating the Diameter of a Drive Shaft when the Torque is Unknown A solid steel drive shaft is to be capable of transmitting 60 hp. at 550 rpm. What should its diameter be if the maximum torsional shear stress is to be kept less than half the tensile yield strength?

Solution Power = P = 60 hp = 745.7 × 60 = 44742 W, N_{rpm} = 550 rev/min, D =?

For steel the tensile yield strength may be taken as: σ_{ys} = 250 MPa (see Table 3.2 & Sect. 9.2).
$\tau < ½ \sigma_{ys} < ½ \times 250 < 125$ MPa, τ = 100 MPa = 10⁸ Pa.
By using the modified form of Eq. 10.8,

$$T = \frac{P}{0.105\, N_{rpm}} = \frac{44742}{0.105 \times 550} = 774.75\, N \cdot m$$

By using Eq. 10.11,

$$D = 1.72 \left(\frac{T}{\tau} \right)^{0.33} = 1.72 \left(\frac{774.75}{10^8} \right)^{0.33} = 1.72 \times 8.98 \times 10^{-2.64} = 15.446 \times 0.00229 = 0.0354\,\mathrm{m}$$

Diameter of shaft = 0.0354 m = 35.4 mm.

Example 10.9 Calculating the Torsional Shear Stress with the Aid of Drawing. Refer to Fig. 10.2, calculate the maximum torsional shear stress induced in the MS shaft.

Solution From Fig. 10.2, L = 1.5 m, D = 0.028 m, $r = \dfrac{D}{2} = 0.014\,\mathrm{m}$, $\theta = 12° = 0.21$ rad, G = 83 GPa.

By using Eq. 10.5,

$$\gamma = \frac{r\theta}{L} = \frac{0.014 \times 0.21}{1.5} = 0.00196\,\mathrm{rad} = 0.00196$$

By using Eq. 3.3,

$$\tau = G\gamma = 83 \times 10^9 \times 0.00196 = 162.7\,MPa$$

The shear stress = 162.7 MPa.

Example 10.10 Calculating the Angle of Twist for Torsion in a Solid Square Shaft A torque of 1000 N · m is acting on a solid square shaft with 40mm × 40mm cross-section and length 1.4 m. The shaft is made in steel with modulus of rigidity of 80 GPa. Calculate the angle of twist in the shaft.

Solution a = 40 mm = 0.040 m, T = 1000 N · m, L = 1.4 m, G = 80 × 10⁹ Pa, θ =?

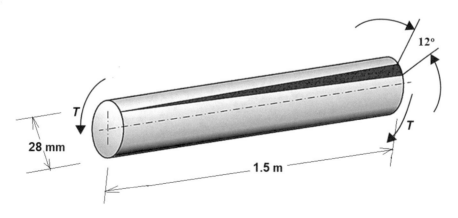

Fig. 10.2 Torsion in shaft with known dimensions and the twist

By using Eq. 10.4.

$$J = \frac{a^4}{6} = \frac{0.04^4}{6} = 426.66 \times 10^{-9} \, m^4$$

By using Eq. 10.7,

$$\theta = \frac{LT}{JG} = \frac{1.4 \times 1000}{426.66 \times 10^{-9} \times 80 \times 10^9} = 0.041 \, radians = 2.35°$$

Example 10.11 Calculating the Torsional Flexibility and the Torsional Stiffness of a Shaft By using the data in Example 10.10, calculate the torsional flexibility and the torsional stiffness of the shaft.

Solution L = 1.4 m, G = 80 × 10⁹ Pa, J = 426.66 × 10⁻⁹ m⁴, f =?, k =?

By using Eqs. 10.13 and 10.14,

$$Torsional \, flexibility = f = \frac{GJ}{L} = \frac{80 \times 10^9 \times 426.66 \times 10^{-9}}{1.4} = 24,380 \, N \cdot m \, / \, rad.$$

$$Torsional \, stiffness = k = \frac{1}{f} = 41 \times 10^{-6} \, rad \, / \, N \cdot m$$

Example 10.12 Determining the Allowable Torque in a Shaft A torque is to act on a solid cylindrical shaft with diameter 60 mm and length 1.5 m. The shaft is made in steel with modulus of rigidity of 80 GPa. The machine design requires the maximum allowable shear stress in the shaft to be 55 MPa. The allowable angle of twist is 3°. Determine the allowable torque in the shaft.

Solution D = 0.060 m, L = 1.5 m, G = 80 × 10⁹ Pa, τ_{all} = 55 × 10⁶ Pa, θ_{all} = 3° = 0.052 rad, T_{all} = ?.

By using Eq. 10.2,

$$J = \frac{\pi}{32} D^4 = \frac{3.142 \times 0.06^4}{32} = 1.27 \times 10^{-6} \, m^4$$

By using the modified form of Eq. 10.1,

$$T_{all} = \frac{\tau_{all} \cdot J}{r} = \frac{55 \times 10^6 \times 1.27 \times 10^{-6}}{0.03} = 2,328 \, N \cdot m$$

By using the modified form of Eq. 10.7,

$$T_{all} = \frac{GJ\theta}{L} = \frac{80 \times 10^9 \times 1.27 \times 10^{-6} \times 0.052}{1.5} = 3,522\,\text{N} \cdot \text{m}$$

In order to avoid failure of the shaft, the minimum torque should be applied *i.e.* T_{all} = 2, 328 N · m.

Example 10.13 Calculating the Diameter of Shaft for Safe Torsional Design A solid cylindrical shaft is designed for torque of 1100 N · m. The shaft is made in steel with modulus of rigidity of 82 GPa. The allowable (maximum) shear stress is 42 MPa, and the allowable angle of twist is 1.2°/m. What should be the diameter of the shaft for the design?

Solution T = 1100 N · m, G = 82 × 10⁹ Pa, τ_{all} = 42 × 10⁶ Pa, θ_{all} = 1.2°/m = 0.021 rad/m, D_{design} = ?.

For the allowable shear stress, we can use Eq. 10.11,

$$D = 1.72 \left(\frac{T}{\tau_{all}}\right)^{0.33} = 1.72 \left(\frac{1100}{42 \times 10^6}\right)^{0.33} = 1.72 \times 26.2^{0.33} \times 0.0105 = 0.053\,\text{m} = 53\,\text{mm}$$

For the allowable angle of twist, we can use the modified form of Eq. 10.7, as follows:

$$J = \frac{LT}{\theta G} = \frac{T}{\theta_{all} G}\left(\text{for 1 m length}, L = 1\right)$$

$$J = \frac{T}{\theta_{all} G} = \frac{1100}{0.021 \times 82 \times 10^9} \times 638.8 \times 10^{-9}\,\text{m}^4$$

By using the modified form of Eq. 10.2,

$$D^4 = \frac{32 J}{\pi} = \frac{32 \times 638.8 \times 10^{-9}}{\pi} = 6505.8 \times 10^{-9}\,\text{m}^4 = 0.065058 \times 10^{-4}\,\text{m}^4$$

$$D = 0.065058^{0.25} \times 10^{-1} = 0.505 \times 10^{-1} = 0.0505\,\text{m} = 50.5\,\text{mm}$$

By comparing the two diameters (53 mm & 50.5 mm), we choose the greater diameter for safe design.

Hence, the shaft should be designed with diameter = 53 mm or practically with D = 55 mm.

Questions and Problems

10.1. Draw a sketch of a shaft subjected to torque, showing the angle of twist, shear strain, and the dimensions.

10.2. Derive an expression for the diameter of a solid shaft in terms of torque and shear stress.

10.3. A torque of 1000 N · m is acting on a hollow cylindrical shaft with length 1.5 m, outer diameter of 40 mm, and inner diameter of 25 mm. The shaft is made in steel with modulus of rigidity of 80 GPa. Calculate the maximum shear stress produced in the shaft.

10.4. A torque of 800 N · m is acting on a solid cylindrical shaft with diameter 40 mm and length 1.4 m. The shaft is made in steel with modulus of rigidity of 82 GPa. Calculate the: (a) maximum shear stress, (b) angle of twist, and (c) shear strain produced in the shaft.

10.5. A 10 kW electric motor shall be used to transmit power through a solid connected shaft. The shaft rotates with 2000 rev/min. The machine design requires the maximum allowable shear stress in the shaft to be 70 MPa. Calculate the torque in the shaft.

10.6. By using the data in *Problem 10.5*, calculate the diameter of the solid cylindrical shaft.

10.7. A solid steel drive shaft is to be capable of transmitting 55 hp. at 500 rpm. What should its diameter be if the maximum torsional shear stress is to be kept less than 2/3 of the tensile yield strength?

10.8. A solid cylindrical shaft is designed for torque of 1200 N · m. The shaft is made in steel with modulus of rigidity of 81 GPa. The allowable (maximum) shear stress is 50 MPa, and the allowable angle of twist is 1.1°/m. What should be the diameter of the shaft?

10.9. A torque of 800 N · m is acting on a solid cylindrical shaft with diameter 40 mm and length 1.4 m. The shaft is made in steel with modulus of rigidity of 82 GPa. Calculate the: (a) torsional flexibility and (b) torsional stiffness.

10.10. A torque is to act on a solid cylindrical shaft with diameter 50 mm and length 1.6 m. The shaft is made in steel with modulus of rigidity of 83 GPa. The machine design requires the maximum allowable shear stress in the shaft to be 60 MPa. The allowable angle of twist is 2°. Determine the allowable torque in the shaft.

References

Ghavami P (2015) Mechanics of materials. Springer Publishing, Berlin, Germany
Hibbeler R (2016) Mechanics of materials. Pearson Education, New York City, USA

Part III
Failure, Design, and Composites Behavior

Chapter 11
Failures Theories and Design

11.1 Failures and Theories of Failure

In Chap. 1, we learnt that a polycrystalline material can fail during monotonic loading in one of the following ways: brittle fracture, ductile fracture, fatigue, creep, hydrogen embrittlement, and stress corrosion cracking (see Sect. 1.3.2, Fig. 1.3). A machine designer must ensure that the component does not fail under the specified (design) loading conditions. This avoidance of failure can be ensured by applying theories of failure. The theories of static failure relate a net effect of the stress state (acting on the body) to the strength of the material. Theories of failure are actually a set of failure criteria developed to predict failure for the ease of design (Christensen, 2013). There are basically two types of mechanical failure: (a) yielding, and (b) fracture. *Yielding* refers to the excessive plastic deformation rendering the machine part unsuitable to perform. *Fracture* involves breaking or tearing apart of the component into two or more parts.

The important failure theories for a material subjected to biaxial stresses include: (a) the maximum principal normal stress theory or *Rankine theory*, (b) maximum shear stress (*MSS*) theory or *Tresca theory*, and (c) von Mises theory. Besides these traditional theories of failure, some modern failure theories have also been developed. Recently, Umantsev has developed a comprehensive theory of plasticity, damage and failure; which includes all major components of the mechanical behavior of ductile materials (*e.g.* strain/work-hardening, Bauschinger effect, etc.) and covers all regimes of viscoplastic tensile/compressive loading/unloading (Umantsev, 2021).

The commonly applied failure theories are explained in the following sections.

Z. Huda, *Mechanical Behavior of Materials*, Mechanical Engineering Series, https://doi.org/10.1007/978-3-030-84927-6_11

11.2   Maximum Principal Normal Stress Theory or *Rankine Theory*

Failure of brittle materials is specified by fracture. The *maximum principal normal stress theory* (or *Rankine theory*) is generally used to predict fracture failure in brittle materials. According to the *maximum principal normal stress theory*, failure occurs when the maximum principal normal stress reaches either the ultimate tensile strength (σ_{ut}) or compressive strength (σ_c). For safe design, the *Rankine theory* is mathematically expressed as (Huda, et al., 2010):

$$-\sigma_c < \{\sigma_1, \sigma_2\} < \sigma_{ut} \tag{11.1}$$

where σ_1 and σ_2 are the principal normal stresses for 2D state of stresses (see Fig. 11.1).

It is evident in Fig. 11.1 and Inequality 11.1 that failure will occur when the maximum principal normal stress induced in the material exceeds the strength of the material. Hence, for safe design, the principal normal stresses must be within the boundaries shown in Fig. 11.1. The application of the maximum principal normal stress theory is illustrated in Example 11.1.

11.3   Maximum Shear Stress Theory of Failure *or Tresca Theory*

11.3.1   *Theoretical Aspect of Tresca Theory*

Failure of ductile materials is specified by the onset of *yielding*. The maximum shear stress (MSS) theory (or *Tresca theory*) is used to predict failure of ductile materials. The MSS theory or *Tresca theory* states that *yielding* would occur when the greatest maximum shear stress ($\tau_{max, crit}$) reaches a critical value – half of the

Fig. 11.1 Graphical representation of the maximum principal normal stress theory

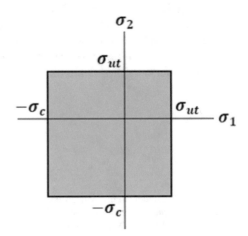

yield strength $\left(\dfrac{\sigma_{ys}}{2}\right)$ of the material. According to the Tresca theory, the failure criterion for yielding can be mathematically expressed as:

$$\tau_{max,\,crit} = \frac{\sigma_{ys}}{2} \tag{11.2}$$

By combining Eq. 11.2 with Eq. 8.18, we obtain:

$$\frac{|\sigma_{max} - \sigma_{min}|}{2} = \frac{\sigma_{ys}}{2} \tag{11.3}$$

or

$$|\sigma_{max} - \sigma_{min}| = \sigma_{ys} \tag{11.4}$$

In the state of plane stress ($\sigma_3 = 0$), the maximum principal normal stress is σ_1. The minimum stress will be σ_3 when σ_2 is positive. In case σ_2 is negative, the minimum stress will be σ_2. The graphical depiction of Tresca theory is given in Fig. 11.2.

Figure 11.2 indicates that in the state of plane stress, failure (yielding) will occur when:

$$|\sigma_1| \geq \sigma_{ys} \tag{11.5a}$$

or

$$|\sigma_2| \geq \sigma_{ys} \tag{11.5b}$$

or

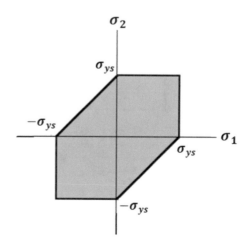

Fig. 11.2 Graphical depiction of Tresca theory or MSS theory

$$\left|\sigma_1 - \sigma_2\right| \ge \sigma_{ys} \tag{11.5c}$$

The significance of Eqs. 11.5a, 11.5b, and 11.5c is illustrated in Examples 11.2 and 11.3.

11.3.2 Design Application of Tresca Theory

In the design of a machine component, it is important to allow a factor of safety (*FoS*). So we can re-write Eq. 11.4 as:

$$\left|\sigma_{max} - \sigma_{min}\right| = \frac{\sigma_{ys}}{\left(FoS\right)} \tag{11.6}$$

In the state of plane stress with positive σ_2, Eq. 11.6 will take the form:

$$\left|\sigma_1 - \sigma_3\right| = \frac{\sigma_{ys}}{\left(FoS\right)} \tag{11.7}$$

Consider the application of Tresca theory to design of a shaft, as shown in Fig. 11.3. By considering the point *A* on the surface of the shaft (Fig. 11.3), $\sigma_y = 0$; so Eq. 8.12 and Eq. 8.13 take the form:

$$\sigma_1 = \frac{\sigma_x}{2} + \frac{\sqrt{\sigma_x^2 + 4\tau_{xy}^2}}{2} \tag{11.8}$$

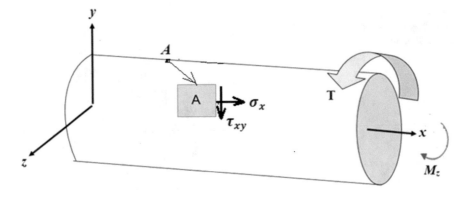

Fig. 11.3 Torque and bending moment acting on a shaft

$$\sigma_3 = \frac{\sigma_x}{2} - \frac{\sqrt{\sigma_x^2 + 4\tau_{xy}^2}}{2} \tag{11.9}$$

By combining Eqs. 11.7, 11.8, and 11.9, we obtain:

$$\sqrt{\sigma_x^2 + 4\tau_{xy}^2} = \frac{\sigma_{ys}}{(FoS)} \tag{11.10}$$

The significance of Eq. 11.10 is illustrated in Example 11.4.

By using the relationships for the bending moment, diameter, and torque of a shaft, we can write:

$$\sigma_x = \frac{32\,M}{\pi\,D^3} \tag{11.11a}$$

$$\tau_{xy} = \frac{16\,T}{\pi\,D^3} \tag{11.11b}$$

where M is the bending moment; T is the torque; and D is the diameter of the shaft.

By combining Eq. 11.10 and 11.11a, 11.11b, we get:

$$\sqrt{\left(\frac{32\,M}{\pi\,D^3}\right)^2 + 4\left(\frac{16\,T}{\pi\,D^3}\right)^2} = \frac{\sigma_{ys}}{(FoS)} \tag{11.12}$$

Equation 11.12 presents the criterion of yielding (failure) of the shaft. In order to ensure safe design, we can re-write Eq. 11.12 as:

$$D^3 \geq \frac{32(FoS)}{\pi\,\sigma_{ys}}\sqrt{M^2 + T^2} \tag{11.13}$$

In order to calculate the factor of safety (*FoS*) for safe design, Eq. 11.13 takes the form:

$$FoS = \frac{\pi\,\sigma_{ys}\,D^3}{32\sqrt{M^2 + T^2}} \geq 1 \tag{11.14}$$

The significance of Eqs. 11.13 and 11.14 is illustrated in Examples 11.5 and 11.6.

11.4   Von Mises Theory of Failure

11.4.1   Theoretical Aspect of von-Mises Theory

The von Mises failure theory states that yielding will occur when the root mean square (RMS) value of the maximum shear stresses reaches a critical value. In the plane stress sheet metal processes, two of the maximum shear stresses have the value $\sigma_{ys}/2$ and the third stress is zero.

According to the von Mises failure theory, yielding of a ductile material will occur when:

$$\sqrt{\frac{\tau_1^2 + \tau_2^2 + \tau_3^2}{3}} = \sqrt{\frac{\left(\dfrac{\sigma_{ys}}{2}\right)^2 + \left(\dfrac{\sigma_{ys}}{2}\right)^2 + 0}{3}} = \sqrt{\frac{2\left(\dfrac{\sigma_{ys}}{2}\right)^2}{3}} \tag{11.15}$$

Multiplying both sides by $\sqrt{6}$, we obtain:

$$\sqrt{2\left(\tau_1^2 + \tau_2^2 + \tau_3^2\right)} = \sigma_{ys} \tag{11.16}$$

By combining Eq. 8.18(a–c) and Eq. 11.16,

$$\sqrt{\frac{1}{2}\left\{(\sigma_1 - \sigma_2)^2 + (\sigma_2 - \sigma_3)^2 + (\sigma_3 - \sigma_1)^2\right\}} = \sigma_{ys} \tag{11.17}$$

For plane stress state, $\sigma_3 = 0$; so Eq. 11.17 takes the form:

$$\sqrt{\sigma_1^2 - \sigma_1\sigma_2 + \sigma_2^2} = \sigma_{ys} \tag{11.18}$$

The significance of Eq. 11.18 is illustrated in Example 11.7.

11.4.2   Design Aspect of von-Mises Theory of Failure

In the preceding sub-section, we specified the criterion of failure (yielding) by Eq. 11.17. In order to ensure safety in design, Eq. 11.17 can be modified by including the factor of safety (FoS). Thus, according to von-Mises failure theory, safety in design is ensured by:

$$\sqrt{\frac{1}{2}\left\{(\sigma_1 - \sigma_2)^2 + (\sigma_2 - \sigma_3)^2 + (\sigma_3 - \sigma_1)^2\right\}} \leq \frac{\sigma_{ys}}{(FoS)} \tag{11.19}$$

For plane stress state, $\sigma_3 = 0$; so Eq. 11.19 takes the form:

$$\sqrt{\sigma_1^2 - \sigma_1\sigma_2 + \sigma_2^2} \leq \frac{\sigma_{ys}}{(FoS)} \tag{11.20}$$

By considering the point A on the surface of the shaft (Fig. 11.3), $\sigma_y = 0$; so Eq. 8.12 was simplified to Eq. 11.8. Similarly, for $\sigma_y = 0$, Eq. 8.13 can be simplified to:

$$\sigma_2 = \frac{\sigma_x}{2} - \frac{\sqrt{\sigma_x^2 + 4\tau_{xy}^2}}{2} \tag{11.21}$$

By combining Eqs. 11.8, 11.20, and 11.21, we get:

$$\sqrt{\left(\frac{\sigma_x}{2}\right)^2 + 3\left(\frac{\sigma_x^2 + 4\tau_{xy}^2}{4}\right)} \leq \frac{\sigma_{ys}}{(FoS)} \tag{11.22}$$

or

$$\sqrt{\sigma_x^2 + 3\tau_{xy}^2} \leq \frac{\sigma_{ys}}{(FoS)} \tag{11.23}$$

By combining Eqs. 11.11a, 11.11b and 11.23, we obtain:

$$\sqrt{\left(\frac{32M}{\pi D^3}\right)^2 + 3\left(\frac{16T}{\pi D^3}\right)^2} \leq \frac{\sigma_{ys}}{(FoS)} \tag{11.24}$$

By solving Eq. 11.24 for the cube of diameter, we get:

$$D^3 \geq \frac{32(FoS)}{\pi\sigma_{ys}}\sqrt{M^2 + \frac{3}{4}T^2} \tag{11.25}$$

The factor of safety can be calculated by re-arranging the terms of Eq. 11.25, as follows:

$$FoS = \frac{\pi\sigma_{ys}D^3}{32\sqrt{M^2 + \frac{3}{4}T^2}} \geq 1 \tag{11.26}$$

The significance of Eqs. 11.25 and 11.26 is illustrated in Examples 11.8 and 11.9.

11.5   Calcualtions – *Worked Examples*

Example 11.1 Predicting Failure in a Brittle Material by Calculating σ_1 and σ_2 A machine component is made of alumina with a tensile strength of 300 MPa. At a point of interest on the free surface of the component, the stresses with respect to a convenient coordinate system in the plane of the surface are $\sigma_x = 210$ MPa, $\sigma_y = 130$ MPa, and $\tau_{xy} = 160$ MPa. Will the failure occur at the specified point?

Solution $\sigma_x = 210$ MPa, $\sigma_y = 130$ MPa, and $\tau_{xy} = 160$ MPa, $\sigma_{ut} = 300$ MPa $\sigma_1 = ?, \sigma_2 = ?$

By using Eqs. 8.12 and 8.13,

$$\sigma_1 = \frac{\sigma_x + \sigma_y}{2} + \frac{\sqrt{\left(\sigma_x - \sigma_y\right)^2 + 4\tau_{xy}^2}}{2} = \frac{210 + 130}{2} + \frac{\sqrt{\left(210 - 130\right)^2 + \left(4 \times 160^2\right)}}{2} = 335\,MPa$$

$$\sigma_2 = \frac{\sigma_x + \sigma_y}{2} - \frac{\sqrt{\left(\sigma_x - \sigma_y\right)^2 + 4\tau_{xy}^2}}{2} = \frac{210 + 130}{2} - \frac{\sqrt{\left(210 - 130\right)^2 + \left(4 \times 160^2\right)}}{2} = 5\,MPa$$

Since the material (alumina) is brittle, we must apply the Rankine theory. Here, $\sigma_1 > \sigma_{ut}$.

Since the maximum principal normal stress ($\sigma_1 = 335$ MPa) exceeds the ultimate tensile strength of the brittle material ($\sigma_{ut} = 300$ MPa), failure will occur.

Example 11.2 Predicting Failure in a Ductile Material when Both Applied Normal Stresses are Tensile A machine component is made of mild steel with a tensile yield strength of 220 MPa. At a point of interest on the free surface of the component, the stresses, in the state of plane stress, with respect to a convenient coordinate system in the plane of the surface are $\sigma_x = 200$ MPa, $\sigma_y = 180$ MPa, and $\tau_{xy} = 15$ MPa. Apply MSS theory to predict failure at the specified point.

Solution By using Eqs. 8.12 and 8.13,

$$\sigma_1 = \frac{\sigma_x + \sigma_y}{2} + \frac{\sqrt{\left(\sigma_x - \sigma_y\right)^2 + 4\tau_{xy}^2}}{2} = \frac{200 + 180}{2} + \frac{\sqrt{\left(200 - 180\right)^2 + \left(4 \times 15^2\right)}}{2}$$

$$\sigma_1 = 190 + \frac{\sqrt{400 + 900}}{2} = 190 + \frac{\sqrt{1300}}{2} = 190 + 18 = 208\,MPa$$

$$\sigma_2 = \frac{\sigma_x + \sigma_y}{2} - \frac{\sqrt{\left(\sigma_x - \sigma_y\right)^2 + 4\tau_{xy}^2}}{2} = \frac{200 + 180}{2} - \frac{\sqrt{\left(200 - 180\right)^2 + \left(4 \times 15^2\right)}}{2}$$

$$\sigma_2 = 190 - 18 = 172\,MPa$$

By applying MSS theory for the ductile material (mild steel) with σ_{ys} = 220 MPa, we can use Eqs. 11.5a, 11.5b, 11.5c, as follow:

$$|\sigma_1| = 208\,MPa < \sigma_{ys},$$

$$|\sigma_2| = |172| = 172\,MPa < \sigma_{ys},$$

$$|\sigma_1 - \sigma_2| = |208 - 172| = 36\,MPa < \sigma_{ys}$$

Since the stresses are below the yield strength, failure will not occur *i.e.* the design is safe.

Example 11.3 Predicting Failure in a Ductile Material when One Applied Normal Stress is Tensile and the Other Compressive A machine component is made of mild steel (σ_{ys} = 220 MPa). At a point of interest on the free surface of the component, the stresses in the plane of the surface are σ_x = 180 MPa, σ_y = 180 MPa, and τ_{xy} = 120 MPa. Predict failure by applying MSS theory.

Solution By using Eqs. 8.12 and 8.13,

$$\sigma_1 = \frac{\sigma_x + \sigma_y}{2} + \frac{\sqrt{(\sigma_x - \sigma_y)^2 + 4\tau_{xy}^2}}{2} = \frac{180 - 200}{2} + \frac{\sqrt{(180 + 200)^2 + (4 \times 120^2)}}{2}$$

$$\sigma_1 = -10 + \frac{\sqrt{144400 + 57600}}{2} = -10 + \frac{\sqrt{202000}}{2} = -10 + 224.7 = 214.7\,MPa$$

$$\sigma_2 = \frac{\sigma_x + \sigma_y}{2} - \frac{\sqrt{(\sigma_x - \sigma_y)^2 + 4\tau_{xy}^2}}{2} = \frac{180 - 200}{2} - \frac{\sqrt{(180 + 200)^2 + (4 \times 120^2)}}{2}$$

$$\sigma_2 = -10 - 224.7 = -234.7\,MPa$$

By using Eqs. 11.5a, 11.5b, 11.5c,

$$|\sigma_1| = 214.7\,MPa > \sigma_{ys},$$

$$|\sigma_2| = |-234.7| = 234.7\,MPa > \sigma_{ys},$$

$$|\sigma_1 - \sigma_2| = |214.7 + 234.7| = 449.4\,MPa > \sigma_{ys}$$

Hence, failure will occur *i.e.* the design is unsafe.

Example 11.4 Predicting Failure by Applying Tresca Theory when $\sigma_y = 0$ A solid shaft is made of mild steel with a tensile yield strength of 220 MPa. At a point of interest on the free surface of the shaft (Fig. 11.3), the stresses, in the state of plane stress, in the plane of the surface are $\sigma_x = 170$ MPa, $\sigma_y = 0$, and $\tau_{xy} = 25$ MPa. Apply MSS (Tresca) theory to predict failure at the specified point, if the factor of safety is 3.

Solution By using Eq. 11.10,

$$\sqrt{\sigma_x^2 + 4\tau_{xy}^2} \leftrightarrow \frac{\sigma_{ys}}{(FoS)}$$

$$\sqrt{170^2 + \left(4 \times 25^2\right)} \leftrightarrow \frac{220}{3}$$

$$177 \leftrightarrow 73$$

$$177 > 73$$

Failure (yielding) will occur *i.e.* the design is unsafe.

Example 11.5 Safely Designing a Shaft by Calculating its Diameter by using *Tresca Theory* A shaft is made of carbon steel with a yield strength of 500 MPa. There are forces on pulleys mounted on the shaft to keep it in equilibrium. The greatest bending moment at pulley A on the shaft is 640×10^3 N - mm, and the torque is 580×10^3 N - mm. Design the shaft by calculating its minimum diameter by using *Tresca theory*. Take the factor of safety as 2.5.

Solution FoS $= 2.5, T = 580 \times 10^3$ N - mm, M $= 640 \times 10^3$ N - mm, $\sigma_{ys} = 500$ MPa, $D = ?$

By using Eq. 11.13 (Tresca theory),

$$D^3 \geq \frac{32(FoS)}{\pi\,\sigma_{ys}} \sqrt{M^2 + T^2} = \frac{32 \times 2.5}{\pi \times 500} \sqrt{640000^2 + 580000^2}$$

$$D^3 = \frac{80}{1571} \sqrt{0.41 \times 10^{12} + 0.336 \times 10^{12}} = 0.051 \times \sqrt{0.746} \times 10^6 = 0.044 \times 10^6 \text{ mm}^3$$

$$D = 0.044^{0.333} \times 10^2 = 0.353 \times 100 = 35.3 \text{ mm}$$

The shaft must be designed with a minimum diameter of 36 mm.

Example 11.6 Calculating the Factor of Safety for Designing a Shaft using **Tresca Theory** A 40-mm-diameter solid shaft is made of carbon steel with a yield strength of 480 MPa. There are forces on pulleys mounted on the shaft to keep it in equilibrium. The greatest bending moment at pulley A on the shaft is 600×10^3 N - mm, and the torque is 530×10^3 N - mm. The shaft showed satisfactory service in the machine. Calculate the factor of safety by using *Tresca Theory*.

Solution $T = 530 \times 10^3$ N - mm, $M = 600 \times 10^3$ N - mm, $\sigma_{ys} = 480$ MPa, $D = 40$ mm, $N =?$

By using Eq. 11.14 (Tresca theory),

$$FoS = \frac{\pi \sigma_{ys} D^3}{32\sqrt{M^2 + T^2}} = \frac{3.142 \times 480 \times 40^3}{32\sqrt{\left(600 \times 10^3\right)^2 + \left(530 \times 10^3\right)^2}} = \frac{96522240}{32\sqrt{0.36 \times 10^{12} + 0.28 \times 10^{12}}}$$

$$FoS = \frac{96522240}{32 \times \sqrt{0.64} \times 10^6} = \frac{96522240}{32 \times 0.8} \times 10^{-6} = 3.77 \geq 3.77$$

The factor of safety should be 3.8 or 4.

Example 11.7 Predicting Failure by Applying Von Mises Failure Theory By using the data in Example 11.2, apply Von Mises failure theory to predict failure.

Solution $\sigma_1 = 208$ MPa, $\sigma_2 = 172$ MPa, $\sigma_{ys} = 220$ MPa

By using Eq. 11.18 (Von Mises failure theory),

$$\sqrt{\sigma_1^2 - \sigma_1\sigma_2 + \sigma_2^2} \leftrightarrow \sigma_{ys}$$

$$\sqrt{208^2 - \left(208 \times 172\right) + 172^2} \leftrightarrow 220\,MPa$$

$$192.5\,MPa < 220\,MPa$$

Hence failure will not occur *i.e.* the design is safe. This proves that Von Mises theory is in agreement with Tresca theory or MSS theory.

Example 11.8 Designing a Shaft by Calculating its Diameter by using *von-Mises Theory* By using the data in Example 11.5, calculate the minimum diameter of the shaft for the safe design by using von-Mises theory.

Solution FoS $= 2.5$, T $= 580 \times 10^3$ N - mm, M $= 640 \times 10^3$ N - mm, $\sigma_{ys} = 500$ MPa, $D =?$

By using Eq. 11.25,

$$D^3 \geq \frac{32\left(FoS\right)}{\pi \sigma_{ys}}\sqrt{M^2 + \frac{3}{4}T^2} \geq \frac{32 \times 2.5}{3.142 \times 500}\sqrt{640000^2 + \frac{3}{4}580000^2}$$

$$D^3 \geq 0.051\sqrt{0.66 \times 10^{12}} = 0.0414 \times 10^6 \text{ mm}^3$$

$$D = 0.0414^{0.333} \times 10^2 = 0.3464 \times 100 = 34.6 \text{ mm} \qquad (11.\text{or})$$

The shaft should be designed with a minimum diameter of 35 mm.

Example 11.9 Calculating Factor of Safety for Designing a Shaft using *Von-Mises Theory* By using the data in Example 11.6, calculate the factor of safety by using von-Mises theory.

Solution T = 530 × 10³ N - mm, *M* = 600 × 10³N - mm, σ_{ys} = 480 MPa, *D* = 40 mm, *N* =?

By using Eq. 11.26,

$$FoS = \frac{\pi \, \sigma_{ys} \, D^3}{32\sqrt{M^2 + \dfrac{3}{4}T^2}} = \frac{3.142 \times 480 \times 40^3}{32\sqrt{600000^2 + \dfrac{3}{4}530000^2}} = \frac{3016320}{\sqrt{0.57}} \times 10^{-6} = 4 \geq 4$$

Questions and Problems

11.1. What is the technological importance of theories of failure?

11.2. Explain the maximum principal normal stress theory with the aid of a diagram.

11.3. Explain Tresca theory giving its mathematical and graphical representations.

11.4. State Von-Mises failure theory and prove that: $\left(\sqrt{1-\alpha+\alpha^2}\right)\sigma_1 = \sigma_{ys}$

11.5. A machine component is made of titanium with a yield strength of 450 MPa. At a point of interest on the free surface of the component, the stresses, in the state of plane stress, with respect to a convenient coordinate system in the plane of the surface are σ_x = 200 MPa, σ_y = 230 MPa, and τ_{xy} = 30 MPa. Apply MSS theory to predict failure at the specified point.

11.6. A machine component is made of cast iron with a tensile strength of 500 MPa. At a point of interest on the free surface of the component, the stresses with respect to a convenient coordinate system in the plane of the surface are σ_x = 330 MPa, σ_y = 250 MPa, and τ_{xy} = 140 MPa. Will the failure occur at the specified point?

11.7. By using the data in 11.5, predict the failure by using the Von-Mises failure theory.

11.8. A machine component is made of aluminum with a yield strength of 35 MPa. At a point of interest on the free surface of the component, the stresses with respect to a convenient coordinate system in the plane of the surface are

$\sigma_x = 100$ MPa, $\sigma_y = -80$ MPa, and $\tau_{xy} = 40$ MPa. Predict failure by applying MSS theory.

11.9. A shaft is made of carbon steel with a yield strength of 470 MPa. There are loads on pulleys mounted on the shaft to keep it in equilibrium. The greatest bending moment at pulley A on the shaft is 620×10^3 N - mm, and the torque is 560×10^3 N - mm. Design the shaft by calculating its minimum diameter by using *Tresca theory*. Take the factor of safety as 3.

11.10. A 45-mm-diameter solid shaft is made of carbon steel with a yield strength of 430 MPa. There are forces on pulleys mounted on the shaft to keep it in equilibrium. The greatest bending moment at a pulley on the shaft is 670×10^3 N - mm, and the torque is 565×10^3 N - mm. The shaft showed satisfactory service in the machine. Calculate the factor of safety by using Von-Mises *Theory*.

11.11. (MCQs). Encircle the most appropriate answers for the following questions.

(a) Which failure theory relates to the root mean square (RMS) value of the maximum shear stresses to a critical value?

(i) Tresca theory, (ii) Rankine theory, (iii) Von-Mises theory, (iv) Umantsev theory

(b) Which failure theory is used for predicting failure of brittle materials?

(i) Tresca theory, (ii) Rankine theory, (iii) Von-Mises theory.

(c) Which failure theory relates the greatest maximum shear stress to $\sigma_{ys}/2$?

(i) Tresca theory, (ii) Rankine theory, (iii) Von-Mises theory, (Iv) Umantsev theory.

(d) The excessive plastic deformation rendering the machine part unsuitable to perform refers to:

(i) Necking, (ii) yielding, (iii) plastic instability, (iv) fracture.

(e) Which failure theory includes all major components of the mechanical behavior of ductile materials (*e.g.* strain/work-hardening, Bauschinger effect, etc.) and covers all regimes of viscoplastic tensile/compressive loading/unloading?

(i) Umantsev theory, (ii) Tresca theory, (iii) Rankine theory, (iv) Von-Mises theory.

References

Christensen RM (2013) Theory of materials failure. Oxford University Press, Oxford, UK

Huda Z, Bulpett R, Lee KY (2010) Design against fracture and failure. Trans Tech Publications, Stafa-Zuerich, Switzerland

Umantsev AR (2021) Bifurcation theory of plasticity, damage and failure. Materials Today Communications 26:2021

Chapter 12
Fracture Mechanics and Design

12.1 Engineering Failures and Evolution of Fracture Mechanics

We established in Chap. 1 that all material failures can be classified into three groups: (a) failures based on material characteristics, (b) failures due to design/manufacturing faults, and (c) failures based on loading/environmental conditions (see Fig. 1.3). In general, materials fail by one of the following mechanisms: ductile failure, brittle fracture, ductile-brittle transition (DBT) failure, fatigue, creep, and the like.

Many reported engineering disasters have been associated with brittle/fast fractures. In particular, the failure of the *Liberty ship* that cracked in half during World War-II is attributed to brittle (fast) fracture. The *Liberty ship* was fabricated by welding steel plates together to form a continuous body. Initially, cracks started at notches or arc-weld spots and subsequently propagated right round the hull; so that finally the ship broke in half. A possible design improvement against this failure could be to incorporate some riveted joints to act as crack arresters in the ship structure. Other examples of historical failures include: the failure of *Titanic* ship, the bridge that failed in Belgium in 1951 only 3 years after its fabrication, and the molasses tank failure in Boston (1919); all these failures are attributed to brittle fractures (Huda et al. 2010; Hertzberg 1996).

Cold environment is one of the causes of failure by ductile-brittle transition (DBT) failure (see Chap. 1, Section 1.3.4). For instance, the failure of the ***Titanic*** ship is attributed to DBT failure that occurred in 1912 causing the death of 1500 passengers. The failure investigations revealed the DBT failure that occurred at $-2°C$ was due to a relatively higher DBT temperature of the material. Another example of DBT failure is the failure of the 25th flight of the space shuttle Challenger, STS-51 L that lasted just 73 seconds on Jan. 28, 1986. The seven crew members aboard Challenger were killed. An investigation into the accident determined that the O-rings used in a joint seal on the rockets were inappropriate for the ambient temperature (2.2 °C) at the time of launch. In the cold, the rings reacted differently to the compressive forces of lift-off (NASA 2005).

The above-mentioned historical engineering failures resulted in the evolution of the field: *fracture mechanics* – a methodology that is used to predict and diagnose failure of a part with an existing crack or flaw. *Fracture mechanics* plays an important role in design against failure of a component with an existing crack or flaw.

12.2   Griffith's Crack Theory

A.A. Griffith initiated the development of *Fracture Mechanics* during World War-I. He studied crack growth in a brittle material and reported that the stress becomes infinity large at the tip of a sharp crack. Griffith showed that some surface energy is required for the growth of a crack; this energy is supplied by the loss of strain energy accompanying the relaxation of local stresses as the crack grows. According to Griffith's crack theory, failure occurs when the surface energy increases to a significantly high value due to the loss of strain energy (Anderson 2017).

Mathematical Modeling Griffith's crack theory can be mathematically modeled by considering a stressed large plate, made of a brittle material, containing a through-thickness central crack of length 2*a* (see Figure 12.1a). The applied stress

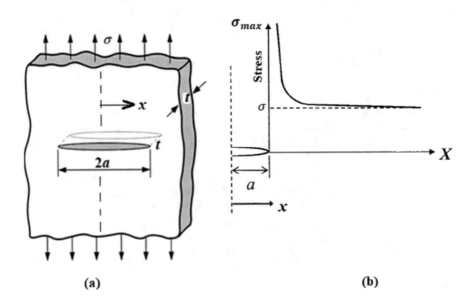

Fig. 12.1 The stressed plate with a through-thickness central crack (a), the stress magnifies at the crack tip (b)

σ is magnified as one moves towards the crack tip; the stress reaches the maximum (σ_{max}) at the tip of crack (see Figure 12.1b). Let the thickness of the plate be t.

The total crack surface area, A, is the sum of the upper and lower surface areas of the crack, *i.e.*

$$A = 2a \cdot t + 2a \cdot t = 4a \cdot t \tag{12.1}$$

The specific surface energy (γ_s) is defined as the surface energy per unit surface area of the crack. Thus the increase in surface energy due to the introduction of the crack (ΔU_1) is given by:

$$\Delta U_1 = 4a \cdot t \cdot \gamma_s \tag{12.2}$$

By using stress analysis for an infinitely large plate containing an elliptical crack, Griffith expressed the decrease in potential energy of the cracked plate (ΔU_2), as follows:

$$\Delta U_2 = -\frac{\pi \cdot \sigma^2 \cdot a^2 \cdot t}{E} \tag{12.3}$$

where E is the Young's modulus of the plate's material. By combining Eqs. 12.2 and 12.3, the net change in potential energy of the plate due to the introduction of the crack will be:

$$U - U_0 = \Delta U_1 + \Delta U_2 = 4a \cdot t \cdot \gamma_s - \frac{\pi \cdot \sigma^2 \cdot a^2 \cdot t}{E} \tag{12.4}$$

where U_0 is the potential energy of the plate without a crack, and U is the potential energy of the plate after the introduction of the crack (see Examples 12.1, 12.2, 12.3, and 12.4).

By re-arranging the terms in Eq. (12.4), we obtain:

$$U = 4a \cdot t \cdot \gamma_s - \frac{\pi \cdot \sigma^2 \cdot a^2 \cdot t}{E} + U_0 \tag{12.5}$$

By taking partial derivatives (with respect to a) for both sides of Eq. 12.5, we get:

$$\frac{\partial U}{\partial a} = 4t \cdot \gamma_s - \frac{2\pi \sigma^2 \cdot a \cdot t}{E} \tag{12.6}$$

At equilibrium, the potential energy of the cracked plate (U) remains constant; by calculus:

$$At\,\text{equilibrium}, \frac{\partial U}{\partial a} = 0 \tag{12.7}$$

By combining Eqs. 12.6 and 12.7, we obtain:

$$\frac{\pi \sigma^2 \cdot a}{E} = 2\gamma_s \tag{12.8}$$

or

$$\sigma = \sqrt{\frac{2E\gamma_s}{\pi a}} \tag{12.9}$$

where σ is the stress applied normal to the major axis of the crack, a is half of the crack length, t is the plate thickness, E is the young's modulus, and γ_s is the specific surface energy of the plate's material. The use of Griffith's crack theory (Eq. 12.9) enables us to calculate the maximum length of the crack that is possible without causing fracture. It must be noted here that the length of a central crack is taken as $2a$, but in case of an edge crack or surface crack, the length of the crack is a (see Example 12.5).

12.3   Stress Concentration Factor

It was explained in the preceding section that the stress applied normal to a crack is magnified as one moves towards a crack tip, and that the stress is the maximum at the tip of crack. The stress concentration factor is a measure of the extent to which the applied stress is magnified at the tip of a crack. The *stress concentration factor* may be defined as the ratio of the maximum stress (σ_m) to the nominal applied tensile stress (σ_0) (see Fig. 12.2).

For static loading, the stress concentration factor can be mathematically expressed in terms of crack geometry as follows:

$$K_s = \frac{\sigma_m}{\sigma_0} = 2\sqrt{\frac{a}{\rho_t}} \tag{12.10}$$

where K_s is the stress concentration factor for static loading; σ_m is the maximum stress at the tip of a central crack, MPa; σ_0 is the stress applied normal to the long axis of the central crack, MPa; a is the one-half the crack length, mm; and ρ_t is the radius of curvature at crack tip, mm. Eq. 12.10 indicates that the *stress concentration factor* (K_s) varies directly as the square root of the crack length, and varies inversely as the square root of the radius of curvature at the crack tip. For a good design of a component with a crack (or a notch), it is important to keep K_s to a minimum. This is why engineering components are designed with a reasonably large radius of curvature (ρ_t). As a case study, we may consider the crash of de Havilland Comet aircraft in 1954. During the Comet flight, the crack of critical size (a_c)

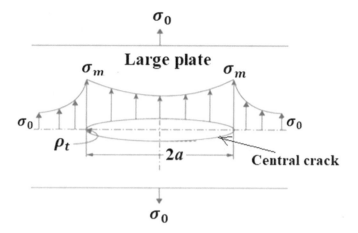

Fig. 12.2 The magnification of applied stress to the maximum stress at the crack tip (ρ_t = radius of curvature at the tip of crack)

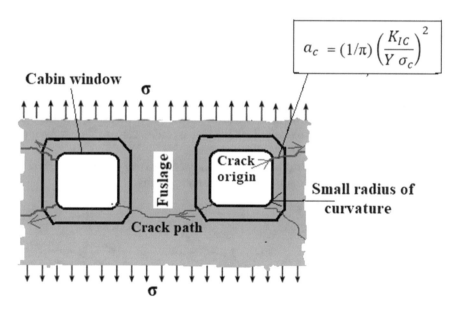

$$a_c = (1/\pi)\left(\frac{K_{IC}}{Y\,\sigma_c}\right)^2$$

Fig. 12.3 The propagation of critical-sized cracks leading to fracture of fuselage due to small radius of curvature in cabin windows in the Comet aircraft

originated from a rivet hole near the cabin window (see Eq. 12.19). When the crack propagated and reached the window, the size of the square-shaped window opening and the small corner radius (ρ) was effectively added to the crack length, leading to fast fracture of the aircraft fuselage (see Fig. 12.3). Thus failure occurred due to the inability of detecting the original crack and due to a small corner radius (ρ) or high stress concentration factor (K_s) in its window (Ashby et al. 2010). This is why, the cabin windows in modern aircrafts are designed with an adequate (reasonably large) corner radii.

By re-arranging the terms in Eq. 12.10, we obtain an expression for ρ_t, as follows:

$$\rho_t = \frac{4a\sigma_0^2}{\sigma_m^2} \tag{12.11}$$

The significance of Eqs. 12.10 and 12.11 is illustrated in Examples 12.6, 12.7, and 12.8.

12.4 Loading Modes in Fracture Mechanics

A machine component can be loaded in any direction relative to crack. However, in many practical situations, there are three loading modes: Mode I, Mode II, and Mode III (Fig. 12.4).

Mode I. Mode I (or *opening mode*) occurs the most often and results in severe damage to the component. In the *opening mode,* a tensile stress acts normal to the plane of the crack. *Mode I* loading may result in normal separation of the crack faces under the action of tensile stresses, which is the most widely encountered in practice.

Mode II. Mode II (or *shearing mode*) causes the second most damage to the component. In the *shearing mode* of loading, a shear stress acts parallel to the plane of the crack thereby tending to slide the crack.

Mode III. Mode III (or *tearing mode*) of loading tends to tear off the crack. It is the least common mode of loading in machine components.

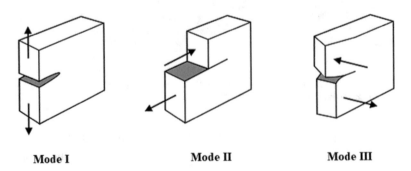

Mode I   Mode II   Mode III

Fig. 12.4 The three modes of loading of cracked components

12.5   Stress Intensity Factor (K), K_c, and K_{IC}

In fracture mechanics, the following three important terms are used to express stress and strength conditions in cracked components: (a) the stress intensity factor, K, (b) the critical stress intensity factor, K_c, and (c) the plane strain fracture toughness, K_{IC}. The three terms can be explained by considering a large plate of infinite width containing a central crack (Figure 12.5a), and a plate with semi-infinite width with an edge crack (Figure 12.5b).

Stress Intensity Factor, K   The *stress intensity factor* refers to the stress level at the crack tip; it depends on the applied stress, crack size, and the crack geometry. Mathematically, the general expression for the stress intensity factor (K) is given by:

$$K = f(\sigma,\, a) = Y\sigma\sqrt{\pi a} \qquad (12.12)$$

where σ is the applied stress, MPa; a is the crack size, m; Y is the geometric factor; and K is the stress intensity factor, $MPa - \sqrt{m}$. By using fracture mechanics analyses, different values of geometric factor Y have been developed for various crack-specimen geometries (see Table 12.1).

In Table 12.1, the term K_I refers to the stress intensity factor in *mode I* loading. The crack geometries corresponding to Eqs. 12.15 and 12.16 are shown in Fig. 12.6. The significance of the data in Table 12.1 and Eqs. 12.13, 12.14, 12.15, and 12.16 is illustrated in Examples 12.9 and 12.10.

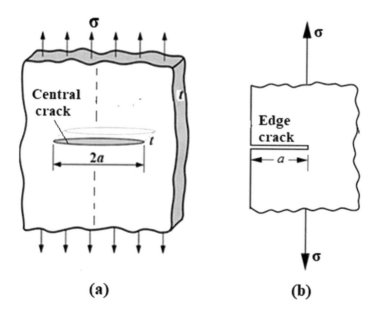

Fig. 12.5 Stressed large plates containing a central crack (a), and an edge crack (b)

Table 12.1 Specimen/crack geometries and the corresponding Y values and K solutions

#	Specimen/crack geometry	Y value	K solution	Equation #
1.	Plate of infinite width with central crack of length $2a$	1.0	$K_I = \sigma\sqrt{\pi a}$	(12.13)
2.	Plate of semi-infinite width an edge crack of length a	1.12	$K_I = 1.12\sigma\sqrt{\pi a}$	(12.14)
3.	Plate of semi-finite width containing an embedded circular flaw of length $2a$	$\dfrac{2}{\neq}$	$K_I = \dfrac{2}{\pi}\sigma\sqrt{\pi a}$	(12.15)
4.	Plate of semi-finite width containing a semi-circular surface flaw of depth a	$1.1\left(\dfrac{2}{\pi}\right)$	$K_I = 1.1\left(\dfrac{2}{\pi}\right)\sigma\sqrt{\pi a}$	(12.16)

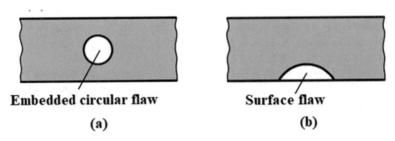

Embedded circular flaw   **Surface flaw**

(a)   **(b)**

Fig. 12.6 Plate of semi-finite width containing embedded circular crack (**a**), and semi-circular surface flaw (**b**).

Critical Stress Intensity Factor, K_c It is evident in Eqs. 12.12, 12.13, 12.14, 12.15, and 12.16 that the stress intensity factor, K, varies with the crack length and the magnitude of applied load. The critical level of stress intensity factor that causes the crack propagation and failure, is called the *critical stress intensity factor*, K_c. Hence, K_c can be expressed as

$$K_c = K \text{ required for crack propagation} = K \text{ for fast fracture} \qquad (12.17)$$

Plain Strain Fracture Toughness K_{IC} *Fracture toughness* is a measure of a material's resistance to fast fracture when there exists a crack. For thick specimens with the thickness much greater than the crack size, the fracture toughness is equal to the critical intensity factor K_c; which can be computed by using Eq. 12.17. But, for thin specimens, the value of K_c strongly depends on the thickness of the specimen. As the specimen-thickness increases, K_c decreases to a constant value; this constant value of K_c, is called the *plane strain fracture toughness*, ***K_{IC}*** (pronounced as kay-one-see). In general, it is the ***K_{IC}*** that is reported as the fracture toughness property of the material since ***K_{IC}*** value does not depend on the thickness of the component. The fracture toughness data ranges for some engineering materials are presented in Table 12.2.

Table 12.2 Fracture toughness (K_{IC}) data for some materials at 25 °C

Material	Fracture toughness (MPa$\sqrt{m}$)	Material	Fracture toughness (MPa$\sqrt{m}$)
Alloy steels	50–87	Polystyrene (PS)	0.7–1.1
Carbon steels	15–90	Polypropylene (PP)	3.0–4.5
Aluminum alloys	24–44	PVC	1.5–4.1
Titanium alloys	50–58	Concrete	0.2–1.4
Copper alloys	35–90	Silica glass	0.7–0.8
Nickel alloys	80–100	Aluminum oxide	2.7–5.0

12.6   Design Philosophy

12.6.1   What Is the Design Philosophy of Fracture Mechanics?

In industrial practice, there are often cases when a structure or a machine component contains a crack. The design philosophy of fracture mechanics enables us to decide whether or not the design of a cracked component is safe for an application. The design philosophy is based on the relationship involving the critical crack size, the critical stress, and K_{IC}, as follows:

$$K \leftrightarrow K_{IC} = Y\sigma_c \sqrt{\pi a_c} \qquad (12.18)$$

where σ_c is the stress to cause fracture – the critical stress, MPa; and a_c is the critical (maximum allowable) crack size, m. If the fracture toughness (K_{IC}) and the critical stress are known, the critical crack size can be calculated by re-arranging the terms in Eq. 12.18, as follows:

$$a_c = \frac{1}{\pi}\left(\frac{K_{IC}}{Y\sigma_c}\right)^2 \qquad (12.19)$$

12.6.2   Application of Design Philosophy to Decide whether or Not a Design Is Safe

Equation 12.18 indicates that there are three variables a_c, σ_c, and K_{IC}; if two variables are known, the third can be determined. For example, if the crack size (a) and the stress (σ) are known, the stress intensity factor (K) can be computed by using Eq. 12.18; the computed K value is then compared with K_{IC} of the material by referring to Table 12.2. If $K < K_{IC}$, the design is safe. On the other hand, if $K \geq K_{IC}$, the design is unsafe (see Examples 12.11 and 12.12).

12.6.3 Application of Design Philosophy to Material Selection

Besides making a decision on the safety of design, Eq. 12.18 enables us to select an appropriate material for a cracked component for an engineering application. For example, if the critical crack size (a_c) and the critical stress (σ_c) values for an application are known, K_{IC} can be computed by using Eq. 12.18; the computed K_{IC} value enables a designer to select the appropriate material having the right fracture toughness (K_{IC}) (by reference to Table 12.2) for the application (see Example 12.13).

12.6.4 Application of Design Philosophy to Design of a Testing/NDT Method

The design philosophy (Eq. 12.18) is also helpful in designing a non-destructive testing (NDT) method to measure the crack size. If a selected material's fracture toughness (K_{IC}) is given and if the design/critical stresses (σ_c) are also specified, we can calculate the critical (maximum) size of the crack (a_c) that can be allowed by using Eq. 12.19. Once the critical crack size is known, it is possible to design an NDT technique that can detect any flaw comparable to the size of computed a_c (see Example 12.14).

12.6.5 Application of Design Philosophy to the Determination of Design Stress

The design philosophy is also helpful in determining the design stress for a stressed application. Once the critical crack size (a_c) has been fixed and the material with a known K_{IC} has been selected, the critical stress can be computed (see Example 12.15).

12.7   Calculations – Worked Examples

Example 12.1 Calculating the Total Surface Area of a Through-Thickness Central Crack A 14-μm long through-thickness central crack is introduced in a 6-mm-thick large plate made of glass. The specific surface energy and Young's modulus of this glass are 0.17 J/m² and 68 GPa, respectively. The plate is under the action of a tensile stress of 35 MPa. Calculate the total surface area of the crack.

Solution $2a = 14 \ \mu m$, $a = 7 \ \mu m = 7 \times 10^{-6}$ m $= 7 \times 10^{-3}$ mm $= 0.007$ mm, $t = 6$ mm.

By using Eq. 12.1,

Total surface area of the crack $= A = 4a \cdot t = 4 \times 0.007$ mm $\times 6$ mm $= 0.168 \ mm^2$

Example 12.2 Calculating the Increase in the Surface Energy of Plate due to a Crack By using the data in Example 12.1, calculate the increase in surface energy of the plate due to the introduction of the crack.

Solution $4a \cdot t = 0.168$ mm² $= 0.168 \ (10^{-3}m)^2 = 0.168 \times 10^{-6} \ m^2$, $\gamma_s = 0.17$ J/m².

By using Eq. 12.2,

$$\Delta U_1 = 4a \cdot t \cdot \gamma_s = 0.168 \times 10^{-6} \times 0.17 = 28.56 \times 10^{-3} \times 10^{-6} = 28.56 \times 10^{-9} J$$

The increase in surface energy of the plate due to the introduction of crack $= \Delta U_1 = 28.56$ nJ.

Example 12.3 Calculating the Decrease in the Potential Energy of a Cracked Plate By using the data in Example 12.1, compute the decrease in potential energy of the cracked plate.

Solution $\sigma = 35$ MPa $= 35 \times 10^6 Pa$, $a = 7 \times 10^{-6}$ m, $t = 6$ mm $= 6 \times 10^{-3}$ m, $E = 68 \times 10^9$ Pa.

By using Eq. 12.3,

$$\Delta U_2 = -\frac{\pi \cdot \sigma^2 \cdot a^2 \cdot t}{E} = -\frac{3.142 \times \left(35 \times 10^6\right)^2 \times \left(7 \times 10^{-6}\right)^2 \times 6 \times 10^{-3}}{68 \times 10^9}$$

$$\Delta U_2 = -\frac{3.142 \times 1225 \times 10^{12} \times 49 \times 10^{-12} \times 6 \times 10^{-3}}{68 \times 10^9} = -16.64 \times 10^{-9} \ J = -16.64 \, nJ$$

Example 12.4 Calculating the Net Change in Energy of Plate Due to Introduction of Crack By using the data in Examples 12.2 and 12.3, calculate the net change in the potential energy of the plate due to the introduction of the crack.

Solution $\Delta U_1 = 28.56$ nJ, $\Delta U_2 = -16.64$ nJ, $U - U_0 = ?$

By uisng Eq. 12.4,

$$U - U_0 = \Delta U_1 + \Delta U_2 = 28.56 + (-16.64) = 28.56 - 16.64 = 11.92 \, nJ \cong 12 \, nJ$$

The net change in the potential energy of the plate due to the introduction of crack $= 12$ nJ.

Example 12.5 Calculating the Fracture Stress by using Griffith Equation A
1-mm-long through-thickness central crack is introduced in a very large plate made
of amorphous silica. The specific surface energy and Young's modulus of silica
4.7 J/m² and 68 GPa, respectively. Calculate the fracture stress.

Solution $2a = 1$ mm, $a = 0.5$ mm $= 0.0005$ m, $\gamma_s = 4.7$ J/m², $E = 68 \times 10^9$ Pa, $\sigma_f = $?.

By using Eq. 12.9,

$$\text{Fracture stress} = \sigma_f = \sqrt{\frac{2E\gamma_s}{\pi a}} = \sqrt{\frac{2 \times 68 \times 10^9 \times 4.7}{\pi \times 0.0005}} = 20 \, MPa$$

**Example 12.6 Computing the Maximum Length of Crack that is possible
Without Fracture** A tensile stress of 45 MPa is applied to a large plate of a glass.
The specific surface energy and Young's modulus of this glass is 0.1 J/m² and
69 GPa, respectively. Determine the maximum length of a through-thickness central
crack that is possible without fracture.

Solution Central crack, $\sigma = 45$ MPa $= 45 \times 10^6$ Pa, $\gamma_s = 0.1$ J/m², $E = 69 \times 10^9$ Pa, $2a = ?$

By the using the re-arranged form of Eq. 12.8,

$$a = \frac{2E\gamma_s}{\pi \sigma^2} = \frac{2 \times 69 \times 10^9 \times 0.1}{3.142 \times \left(45 \times 10^6\right)^2} = \frac{13.8 \times 10^9}{6.36 \times 10^{15}} = 2.17 \times 10^{-6} \, m = 2.17 \mu m$$

$$2a = 4.34 \, \mu m$$

The maximum length of the crack that is possible without fracture = 4.0 μm.

Example 12.7 Calculating the Maximum Stress at the Crack Tip A large
ceramic plate contains a 0.15-mm-long central crack with a radius of curvature of
0.48 μm at the crack tip. A tensile stress of 42 MPa is applied normal to the major
axis of the crack. Calculate the maximum stress at the crack tip.

Solution $2a = 0.15$ mm, $a = 0.75$ mm, $\rho_t = 0.48 \times 10^{-6}$ m $= 0.48 \times 10^{-3}$ mm,
$\sigma_0 = 42$ MPa.

By using the re-arranged form of Eq. 12.10,

$$\text{The maximum stress at the crack tip} = \sigma_m = 2 \, \sigma_0 \sqrt{\frac{a}{\rho_t}} = 2 \times 42 \times \sqrt{\frac{0.75}{0.48 \times 10^{-3}}} = 3317 \, MPa$$

**Example 12.8 Calculating the Stress Concentration Factor for Static
Loading.** By using the data in Example 12.7, calculate the stress concentration
factor for static loading.

Solution By using Eq. 12.10,

$$\text{The stress concentration factor for static loading} = K_s = \frac{\sigma_m}{\sigma_0} = \frac{3317\,MPa}{42\,MPa} = 79$$

Example 12.9 Calculating the radius of curvature at the crack tip An advanced flaw-free ceramic material has a tensile strength of 418 MPa. An elliptical thin 0.3-mm-deep crack is observed before the ceramic plate is tensile tested. The plate fails by brittle fracture at a normal stress of 12 MPa. Calculate the radius of curvature at the crack tip.

Solution $\sigma_{ut} = \sigma_m = 418$ MPa, applied normal stress $= \sigma_0 = 12$ MPa, $a = 0.3$ mm.

By using Eq. 12.11,

$$\rho_t = \frac{4a\sigma_0^2}{\sigma_m^2} = \frac{4 \times 0.3 \times 12^2}{418^2} = \frac{172.8}{174724} = 989 \times 10^{-6}\ mm = 0.989 \times 10^{-6}\ m = 0.989\,\mu m$$

The radius of curvature at the crack tip = 0.989 μm.

Example 12.10 Calculating the stress intensity factor for an Embedded Circular Flaw A plate of semi-finite width contains an embedded circular flaw of length 12 mm. A tensile stress of 650 MPa is applied to the plate in *mode I*. Calculate the stress intensity factor.

Solution Embedded circular crack; $2a = 12$ mm $= 0.012$ m, $a = 0.006$ m, $\sigma = 650$ MPa, $K_I = ?$

By using Eq. 12.15,

$$K_I = \frac{2}{\pi}\sigma\sqrt{\pi a} = \frac{2}{\pi} \times 650 \times \sqrt{3.142 \times 0.006} = 413.74 \times \sqrt{0.0188} = 56.73\,MPa\sqrt{m}$$

Example 12.11 Calculating the Stress Intensity Factor for an Edge Crack in a Plate A plate of semi-infinite width contains an edge crack of length 5 mm. A tensile stress of 400 MPa is applied to the plate in *mode I*. Calculate the stress intensity factor.

Solution Edge crack; $a = 5$ mm $= 0.005$ m, $\sigma = 400$ MPa.

By using Eq. 12.14,

$$K_I = 1.12\sigma\sqrt{\pi a} = 1.12 \times 400 \times \sqrt{3.142 \times 0.005} = 448 \times \sqrt{0.0157} = 56.13\,MPa\sqrt{m}$$

Example 12.12 Deciding Safety of Design for a Very Wide Plate with a Central Crack A 3-mm long central crack was detected in an infinitely wide plate made of aluminum alloy having a plain strain fracture toughness of $30\,MPa\sqrt{m}$. Is this plate's design safe to use for an application where design stresses are 600 MPa?

Solution Central crack; $2a$ = 3 mm = 0.003 m, a = 0.0015 m; $\sigma = 600$ MPa; $K_{IC} = 30\,MPa\sqrt{m}$.

By using Eq. 12.13,

$$K_I = \sigma\sqrt{\pi a} = 600\sqrt{\pi \times 0.0015} = 600 \times \sqrt{47 \times 10^{-4}} = 600 \times 0.01 \times \sqrt{47} = 41.1\,MPa$$
$$K = 41.1\,MPa\sqrt{m}, K_{IC} = 30\,MPa\sqrt{m}$$

Since K > K_{IC}, the design is unsafe.

Example 12.13 Deciding Safety of a Design with a Semi-Circular Surface Flaw in a Plate A plate of semi-finite width contains a semi-circular surface flaw of depth 2 mm. The plain strain fracture toughness of the pate's material is $40\,MPa\sqrt{m}$. Is the plate's design safe for an application with a design stress of 300 MPa?

Solution Semi-circular surface flaw; a = 2 mm = 0.002 m, $\sigma = 300$ MPa; $K_{IC} = 40\,MPa\sqrt{m}$.

By using Eq. 12.16,

$$K_I = 1.1\left(\frac{2}{\pi}\right)\sigma\sqrt{\pi a} = 1.1 \times 0.636 \times 300 \times \sqrt{3.142 \times 0.002} = 16.6\,MPa$$

Since K < K_{IC}, the design is safe.

Example 12.14 Selecting a Material by Applying Design Philosophy of Fracture Mechanics Select a specific ceramic material for a plate of an infinite width with a central crack of length 3 mm; which fails at a stress of 48 MPa.

Solution $2a_c$ = 3 mm = 0.003 mm, a_c = 0.0015 m, σ_c = 48 MPa.

By using Eq. 12.13,

$$K_{IC} = \sigma_c\sqrt{\pi a_c} = 48 \times \sqrt{\pi \times 0.0015} = 3.3\,MPa\sqrt{m}$$

The ceramic material to be selected will be the one having $K_{IC} > 3.3\,MPa\sqrt{m}$.

By reference to Table 12.2, we must select aluminum oxide for this design application.

Example 12.15 Designing an NDT Method by using the Design Philosophy An aluminum alloy plate of semi-finite width has a plane-strain fracture toughness of $40\,MPa\sqrt{m}$; the plate is subjected to a stress of 250 MPa during service. Design an

NDT method capable of detecting an edge crack before the likely growth of the crack.

Solution $K_{IC} = 40\,MPa\sqrt{m}$, $\sigma_c = 250$ MPa, Edge crack, Y = **1.12**, $a_c = ?$.

By using Eq. 12.19,

$$a_c = \frac{1}{\pi}\left(\frac{K_{IC}}{Y\sigma_c}\right)^2 = \frac{1}{\pi}\left(\frac{K_{IC}}{Y\sigma_c}\right)^2 = 0.3183 \times \left(\frac{40}{1.12 \times 250}\right)^2 = 0.0065\,m = 6.5\,mm$$

$$\text{The critical crack size} = a_c = 6.5\,mm$$

A 6.5-mm-long crack can be visually detected. However, it is safer to conduct dye penetrant test.

Example 12.16 Determining the Critical Stress by Applying the Design Philosophy A plate of semi-finite width contains an embedded circular flaw of length 2.8 mm. The plate's material has a fracture toughness of $35\,MPa\sqrt{m}$. What design (working) stresses do you recommend for this application under *mode I* loading?

Solution Embedded circular flaw; $2a = 2.8$ mm $= 2.8 \times 10^{-3}$ m, $a = 0.0014$ m; $K_{IC} = 35\,MPa\sqrt{m}$.

By using Eqs. 12.15 and 12.18,

$$K_{IC} = \frac{2}{\pi}\sigma_c\sqrt{\pi a_c}$$

12. or

$$\sigma_c = \frac{K_{IC}}{2}\sqrt{\frac{\pi}{a_c}} = \frac{35}{2}\sqrt{\frac{3.142}{0.0014}} = 17.5 \times \sqrt{2244.3} = 829\,MPa$$

Since the critical stress is 829 MPa, the design (working) stress should be around 800 MPa.

Questions and Problems

12.1. (MCQs). Encircle the correct answer for each of the following questions.

(a) Which loading mode occurs the most often and results in severe damage to the component?

(i) Mode I, (ii) Mode II, (iii) Mode III, (iv) Mode IV.

(b) Which loading mode is called the tearing mode?

(i) Mode I, (ii) Mode II, (iii) Mode III, (iv) Mode IV.

(c) Which loading mode is called the sliding mode?

(i) Mode I, (ii) Mode II, (iii) Mode III, (iv) Mode IV.

(d) What is the value of geometric factor for a plate with embedded circular crack?

(i) $Y = 1$, (ii) $Y = 1.12$, (iii) $Y = \dfrac{2}{\pi}$, (iv) $Y = 1.1\left(\dfrac{2}{\pi}\right)$.

(e) What is the value of geometric factor for a plate of infinite width with a central crack?

(i) $Y = 1$, (ii) $Y = 1.12$, (iii) $Y = \dfrac{2}{\pi}$, (iv) $Y = 1.1\left(\dfrac{2}{\pi}\right)$.

(f) What is true for a plate with semi-finite width containing a semi-circular surface flaw?

(i) $Y = 1$, (ii) $Y = 1.12$, (iii) $Y = \dfrac{2}{\pi}$, (iv) $Y = 1.1\left(\dfrac{2}{\pi}\right)$.

(g) What is the value of geometric factor for a plate of semi-infinite width with an edge crack?

(i) $Y = 1$, (ii) $Y = 1.12$, (iii) $Y = \dfrac{2}{\pi}$, (iv) $Y = 1.1\left(\dfrac{2}{\pi}\right)$.

(h) Which term is used to express K required for crack propagation?

(i) stress intensity factor, (ii) critical stress intensity factor, (iii) toughness.

(i) Which term is used to express a material's resistance to fast fracture in presence of crack?

(i) stress intensity factor, (ii) critical stress intensity factor, (iii) fracture toughness.

12.2. (a) Explain the role of stress concentration in design of machine components.
(b) Why are the cabin windows of modern aircrafts designed with a large corner radii?

12.3. Derive the mathematical relationship representing the Griffith's crack theory.

12.4. A tensile stress of 70 MPa is applied normal to the major axis of a surface flaw in a large plate of a glass ceramic. The Young's modulus and the specific surface energy for the glass are 70 GPa and 0.27 Jm^{-2}, respectively. Compute the maximum length of the crack that is allowable without causing fracture.

12.5. A large ceramic plate contains a 0.2-mm-long central crack with a radius of curvature of 0.6 μm at the Crack Tip. A Tensile Stress of 38 MPa Is Applied Normal to the Major Axis of the crack. Calculate the stress concentration factor for the static loading.

12.6. A 12-μm long through-thickness central crack is introduced in a 7-mm-thick large plate made of glass having a specific surface energy of 0.18 J/m^2, and

Young's modulus of 67 GPa. A tensile stress of 33 MPa is applied normal to the major axis of the crack. Calculate the net change in the potential energy of the plate due to the introduction of the crack.

12.7. A plate of infinite width contains a central flaw of length 3.2 mm. The plate's material has a fracture toughness of $43\,MPa\sqrt{m}$. What design (working) stresses do you recommend for this application under *mode I* loading?

12.8. A metal plate of semi-finite width has a plane-strain fracture toughness of $30\,MPa\sqrt{m}$; the plate is subjected to a stress of 220 MPa during service. Design an NDT method capable of detecting an edge crack before the likely growth of the crack.

12.9. A plate of semi-finite width contains an embedded circular flaw of length 3 mm. The plain strain fracture toughness of the pate's material is $50\,MPa\sqrt{m}$. Is the plate's design safe for an application with a design stress of 200 MPa?

References

Anderson TL (2017) Fracture mechanics: fundamentals and applications, 4th edn. CRC Press, Boca Raton, FL, USA

Ashby M, Shercliff H, Cebon D (2010) Materials: engineering, science, and design (2nd edition). Butterworth-Heinemann/Elsevier, Amsterdam

Huda Z, Bulpett R, Lee KY (2010) Design against fracture and failure. Trans Tech Publications, Switzerland

Hertzberg RW (1996) Deformation and fracture mechanics of engineering materials. John Wiley & Sons, New York, USA

NASA (2005), NASA-STS-51L Mission profile, https://www.nasa.gov/mission_pages/shuttle/shuttlemissions/archives/sts-51L.html Accessed on December 23, 2020

Chapter 13
Fatigue Behavior of Materials

13.1 Fatigue Failure – *Fundamentals*

When a component is subjected to cyclic stresses over a long period of time, it may fail after a certain number of cycles even though the maximum stress in any cycle is considerably less than the breaking strength of the material. This failure of material, is called *fatigue*. Examples of components that usually fail by fatigue include: rotating shafts, aircraft wings, springs, turbine blades, gears, automobile shock absorbers, and the like.

A fatigue failure generally occurs in three stages: (I) *crack initiation*, the crack originates at a point of stress concentration or a metallurgical flaw (e.g. inclusion); (II) *crack propagation*, the crack propagates across the part under cyclic stresses, and (III) *crack termination*. Figure 13.1 shows a broken stub of a secondary shaft; which failed by fatigue. The decolored part shows the crack initiation stage. Beach marks indicate crack propagation across the failed section, and the jagged region represents the final failure or crack termination (**Schijve** 2009).

13.2 Stress Cycles

13.2.1 *Types of Stresses and Stress Cycles*

Machine components are generally subjected to stresses in one of the following three main ways: (a) axial stress, (b) bending or flexural stress, and (c) torsional stress (see Fig. 13.2). Likewise, there are three types of stress cycle with which loads may be applied to a component: (a) alternating stress cycle, (b) repeating stress cycle, and (c) fluctuating stress cycle (see Fig. 13.3). The simplest type of stress cycle is the alternating stress cycle; which has a symmetrical amplitude about

Z. Huda, *Mechanical Behavior of Materials*, Mechanical Engineering Series, https://doi.org/10.1007/978-3-030-84927-6_13

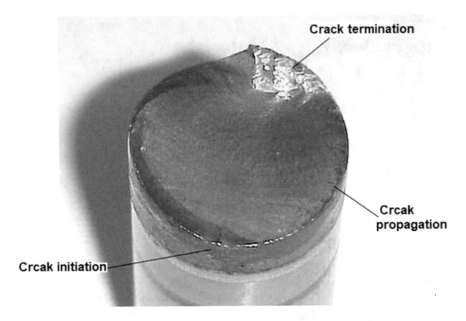

Fig. 13.1 Fatigue failure stages in a stub of a secondary shaft. (*Courtesy* – Reprinted by Permission of Jack Kane, CEO, *EPI Inc.*, USA)

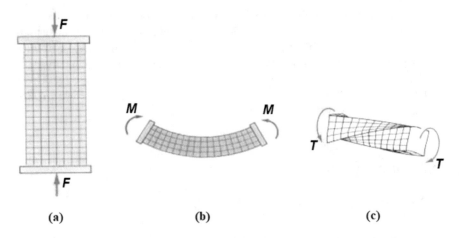

Fig. 13.2 Three types of stresses; (a) axial stress, (b) bending stress, and (c) torsional stress

the *x*-axis. In the alternating stress cycle, the maximum and minimum stresses are equal, but opposite in sign, as in the case of a sine wave (Fig. 13.3a). For example, the stress cycle in an axle is *alternating*; here after every half turn (or half cycle), the stress on a point would be reversed.

In *repeated stress cycle*, the maximum stress (σ_{max}) and the minimum stress (σ_{min}) are asymmetric (Fig. 13.3b). An example of repeated stress cycle is the stress cycle

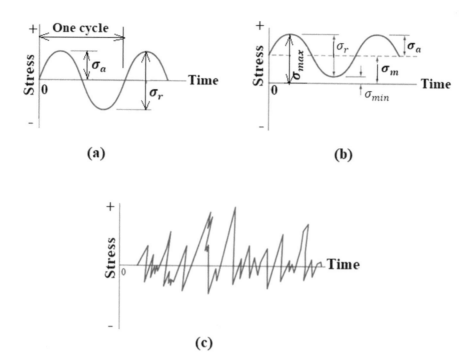

Fig. 13.3 Types of stress cycles: (a) alternating stress cycle, (b) repeating stress cycle, and (c) fluctuating stress cycle

in cantilever beam. In *fluctuating stress cycle*; the stress and frequency vary randomly (Fig. 13.3c); an example is automobile shock absorbers.

13.2.2   Stress Cycle Parameters

In components subjected to cyclic loading, it is important to determine the various stress cycle parameters; which include: (a) mean stress, (b) stress range, (c) stress amplitude, (d) stress ratio, and (e) amplitude ratio. The *mean stress* (σ_m) is the average of maximum and minimum stresses in a cycle. Mathematically,

$$\sigma_m = \frac{\sigma_{max} + \sigma_{min}}{2} \tag{13.1}$$

where σ_{max} is the maximum stress; and σ_{min} is the minimum stress. In case of alternating stress cycle, the mean stress is zero, 0 (see Fig. 13.3a) (see Example 13.1).

The stress range (σ_r) is the difference of the maximum stress and the minimum stress.

$$\sigma_r = \sigma_{max} - \sigma_{min} \qquad (13.2)$$

The stress amplitude (σ_a) in the alternating stress cycle is one-half of the stress range.

$$\sigma_a = \frac{\sigma_r}{2} \qquad (13.3)$$

The stress ratio (R_s) is the ratio of the minimum stress to the maximum stress. Mathematically,

$$R_s = \frac{\sigma_{min}}{\sigma_{max}} \qquad (13.4)$$

The *amplitude ratio* (R_a) is the ratio of the stress amplitude to the mean stress. Mathematically,

$$R_a = \frac{\sigma_a}{\sigma_m} \qquad (13.5)$$

where σ_a is the stress amplitude, and σ_m is the mean stress. The significance of Eqs. 13.2, 13.3a, 13.3b, 13.4, and 13.5 is illustrated in Examples 13.2, 13.3, and 13.4.

13.3   Fatigue Testing – *Determination of Fatigue Strength and Fatigue Life*

Fatigue testing is a type of mechanical testing that is performed by applying cyclic loading to a test piece (Huda et al. 2020). In fatigue testing, a cylindrical test piece is subjected to compressive and tensile stresses as the test-piece is simultaneously bent and rotated by use of a motor and bearings (see Fig. 13.4). The results, from the fatigue test, are plotted in the form of an S-N curve (stress amplitude versus number of cycles to failure curve) (see Fig. 13.5).

The S-N plot in Fig. 13.5 indicates that steel exhibits a fatigue limit or an endurance limit (S_e), whereas an aluminum alloy does not. *Endurance limit* or *fatigue limit* (S_e) is the stress level below which a test material can withstand cyclic stresses indefinitely without exhibiting *fatigue* failure. Most carbon steels and low-alloy steels exhibit a definite fatigue limit (S_e); which is generally about one-half of the tensile strength (σ_{ut}) *i.e.*

$$S_e = \frac{\sigma_{ut}}{2} \qquad (13.6)$$

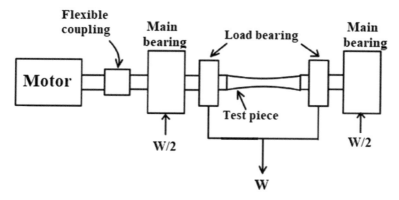

Fig. 13.4 Schematic of rotating-bending fatigue testing machine

Fig. 13.5 S-N Plots for steel and an aluminum alloy (S_e = endurance limit)

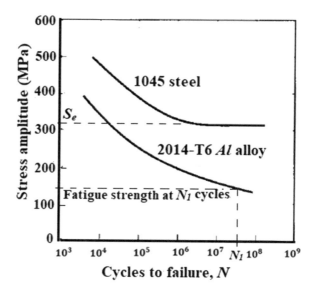

The significance of Fig. 13.4 and Eq. 13.6 is illustrated in Examples 13.5.

It is important to distinguish between two important terms: *fatigue strength* and *fatigue life*. *Fatigue strength* is the stress amplitude at which failure occurs for a given number of cycles. *Fatigue life* (N_f) is the number of cycles to failure at a certain stress (see Examples 13.5 and 13.6). *High cycle fatigue* (*HCF*) refers to long fatigue life; which is typically in the range of 10^3–10^7 cycles. Examples of components that experience *HCF* include: aircraft wings, turbine blades, rotating shafts, and the like.

13.4 Goodman's Law

An empirical relationship between stress amplitude, mean stresses, tensile strength, and fatigue strength can be expressed by the Goodman's law, as follows:

$$\frac{\sigma_a}{S_e} + \frac{\sigma_m}{\sigma_{ut}} = 1 \tag{13.7}$$

where σ_a is the stress amplitude, S_e is the fatigue limit, σ_m is the mean stress, and σ_{ut} is the ultimate tensile strength (see Example 13.7). The Goodman law can be represented by a simple Goodman's diagram; which indicates the limits of safe stresses and failure stresses (Fig. 13.6).

13.5 Techniques in Designing against Fatigue Failure

The major techniques for designing against fatigue failure include: (a) improving surface conditions, (b) avoiding notches/stress-concentration sites in the component design, (c) controlling microstructural defects and grain size, (d) corrosion control, (e) minimizing manufacturing defects (*e.g.* initial flaws, machining marks, etc), (f) maximizing the critical crack size, and (g) introducing compressive residual stresses (Huda and Bulpett 2012).

A good surface condition is very important to avoid fatigue failure because fatigue cracks generally initiate at a defective surface; and then propagate through the bulk material. Since the surface defects are the regions of stress concentrations,

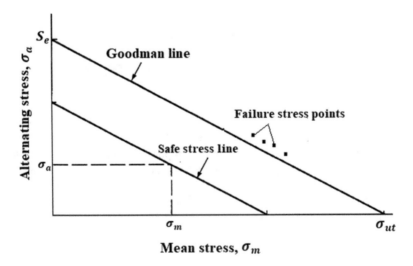

Fig. 13.6 Goodman's diagram for alternating stresses

it is important that the surface must be smooth and free from surface defects (such as scratches, machining marks, fillets, dents, etc.). Additionally, any notch or sharp corner (stress concentration sites) in the component design must be avoided. The fatigue stress concentration factor is a function of flaw geometry as well as the material and the type of loading. The material aspect of the fatigue stress concentration factor is generally expressed as a notch sensitivity factor (q_n), which is defined as:

$$q_n = \frac{K_f - 1}{K_s - 1}$$ (13.8)

where K_s is the static stress concentration factor; and K_f is the fatigue stress concentration factor.

$$K_f = \frac{Fatigue\ limit\ for\ notch - free\ specimen}{Fatigue\ limit\ for\ notched\ specimen}$$ (13.9)

Equation 13.9 indicates that the fatigue strength of a component is reduced by a factor K_f by the effect of notch in the component design (see Examples 13.8 and 13.9). It is evident in Eq. 13.8 that the range of q_n lies between 0 (when $K_f = 1$) and 1 (when $K_f = K_s$).

13.6   Miner's Law of Cumulative Damage

Many structures (e.g. bridges, aircraft fuselage, etc.) are subjected to a range of stress fluctuations, mean levels, and frequencies. It is, therefore, important to predict the fatigue life of a component on the basis of cumulative damage. Miner's theory of cumulative damage suggests that the fatigue damage introduced by a given stress level is proportional to the *cycle ratio i.e.* the number of applied cycles (n) at a stress level divided by the number of cycles to failure (N) at the stress level. Let the fatigue life at a stress level σ_1 be N_1 cycles; and the fatigue life at another stress level σ_2 be N_2 cycles. According to Miner's law,

$$\frac{n_1}{N_1} + \frac{n_2}{N_2} = 1$$ (13.10)

where n_1 is the number of applied cycles at the stress level σ_1, and n_2 is the number of applied cycles at the stress level σ_2. Miner's law is graphically illustrated in Fig. 13.7.

By re-arranging the terms in Eq. 13.10, we obtain:

$$n_2 = N_2 \left(1 - \frac{n_1}{N_1} \right)$$ (13.11)

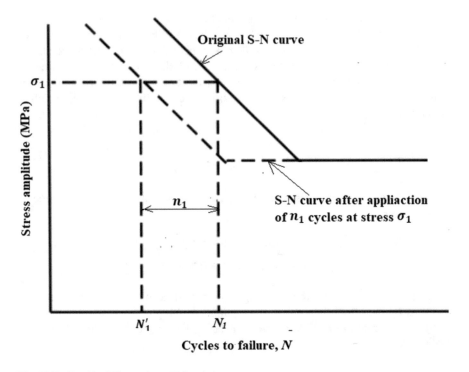

Fig. 13.7 Graphical illustration of Miner's law

Equation 13.10 can be generalized to obtain the generalized Miner's law, as follows:

$$\frac{n_1}{N_1} + \frac{n_2}{N_2} + \frac{n_3}{N_3} + \cdots + \frac{n_i}{N_i} = 1 \qquad (13.12)$$

where n_i is the number of cycles applied at a load corresponding to a fatigue life of N_i cycles.

The significance of Miner's law is illustrated in Examples 13.10, 13.11, 13.12, and 13.13.

13.7   Fatigue Crack Growth Rate and Computation of Fatigue Life

In engineering structures (especially in aircrafts), it is technologically important to predict rate of crack growth during load cycling. For a pre-existing cracked structure or component, the crack growth rate (da/dN) is a function of both stress σ and

Fig. 13.8 Dependence of fatigue crack growth rate on stress and crack size

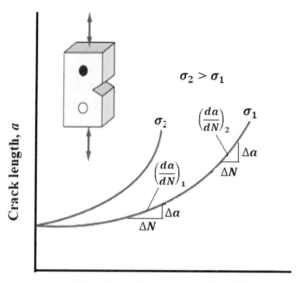

crack size a (see Fig. 13.8). Since in cyclic loading the stress ranges from σ_{min} to σ_{max}, we may use Eq. 12.12 to expressed the stress intensity factor (K), as follows:

$$K_{min} = Y\sigma_{min}\sqrt{\pi a} \qquad (13.13a)$$

$$K_{max} = Y\sigma_{max}\sqrt{\pi a} \qquad (13.13b)$$

So

$$\Delta K = K_{max} - K_{min} = Y\left(\Delta\sigma\right)\sqrt{\pi a} \qquad (13.14)$$

It has been experimentally proved that crack growth rate (da/dN) can be co-related with the cyclic variation in stress intensity factor (ΔK) by Paris law, as follows (Rajabipour et al. 2015):

$$\frac{da}{dN} = A\left(\Delta K\right)^m \qquad (13.15)$$

where A and m are constants; and m is in the range of 3–7. By re-writing Eq. 13.15,

$$dN = \frac{da}{A\left(\Delta K\right)^m} = \frac{da}{A\left(Y\left(\Delta\sigma\right)\sqrt{\pi a}\right)^m} \qquad (13.16)$$

$$\text{or } \int_{0}^{N_f} dN = N_f = \int_{a_0}^{a_f} \frac{da}{A \left(Y \left(\Delta\sigma \right) \sqrt{\pi a} \right)^m} \quad (13.17).$$

where N_f is the fatigue life (number of cycles to failure) of the component subjected to the stress range $\Delta\sigma$ (MPa); a is the crack size, m; Y is the geometric factor; and a_0 the initial crack size, m; and a_f is the final crack size, m; (see Examples 13.14, 13.15, 13.16, 13.17, and 13.18).

13.8   Calculations – *Worked Examples*

Example 13.1 Determining the Mean Stress and Identifying the Type of Stress Cycle The bending stresses acting on a high swaying chimney were determined to be ±78 MPa. Calculate the mean stress. Which type of stress cycle is acting on the chimney?

Solution $\sigma_{max} = 78$ MPa, $\sigma_{min} = -78$ MPa, $\sigma_m = ?$.

By using Eq. 13.1,

$$\text{Mean stress} = \sigma_m = \frac{\sigma_{max} + \sigma_{min}}{2} = \frac{78 - 78}{2} = 0$$

Since the mean stress is zero, the stress cycle acting on the chimney is *alternating stress cycle*.

Example 13.2 Calculating the Stress Range and the Stress Amplitude By using the data in Example 13.1, calculate the stress range and stress amplitude.

Solution   By using Eqs. 13.2 and 13.3a, 13.3b,

$$\text{Stress range} = \sigma_r = \sigma_{max} - \sigma_{min} = 78 - (-78) = 78 + 78 = 156 \, \text{MPa}$$

$$\text{Stress amplitude} = \sigma_a = \frac{\sigma_r}{2} = \frac{156}{2} = 78 \, \text{MPa}$$

Example 13.3 Calculating the Stress Ratio and the Amplitude Ratio By using the data in Examples 13.1 and 13.2, calculate the stress ratio and the amplitude ratio.

Solution $\sigma_{max} = 78$ MPa, $\sigma_{min} = -78$ MPa, $\sigma_m = 0$, $\sigma_a = 78$ MPa.

By using Eqs. 13.4 and 13.5,

$$\text{Stress ratio} = R_s = \frac{\sigma_{min}}{\sigma_{max}} = \frac{-78\,\text{MPa}}{78\,MPa} = -1$$

$$\text{Amplitude ratio} = R_a = \frac{\sigma_a}{\sigma_m} = \frac{78}{0} = \infty$$

Example 13.4 Calculating the Maximum and the Minimum Stresses in a Stress Cycle A component is subjected to an alternating stress cycle with a stress range of 27 MPa. Calculate the maximum and the minimum stresses for the stress cycle.

Solution $\sigma_r = 27$ MPa, $\sigma_m = 0$ (see Sect. 13.2), $\sigma_{max} = ?$, $\sigma_{min} = ?$.

By using Eqs. 13.1 and 13.2,

$$\sigma_{max} - \sigma_{min} = 27 \tag{13.17a}$$

$$\frac{\sigma_{max} + \sigma_{min}}{2} = 0 \tag{13.17b}$$

By solving Eq. 13.17a and 13.17b,
$\sigma_{max} = 13.5$ MPa, $\sigma_{min} = -13.5$ MPa.

Example 13.5 Determining the Tensile Strength of Steel with Reference to S-N Curve By reference to the S-N curve in Fig. 13.5, estimate the tensile strength of the tested steel.

Solution By reference to Fig. 13.5, the fatigue limit (S_e) of the tested steel is around 320 MPa.

By using Eq. 13.6,

$$\text{Tensile strength} = \sigma_{ut} = 2 \times S_e = 2 \times 320 = 640\,\text{MPa}$$

Example 13.6 Determining the Fatigue Strength and Fatigue Life by using S-N Curve By reference to Fig. 13.5, determine the: (a) fatigue strength of the aluminum alloy at 10^7 cycles, (b) fatigue life of the aluminum alloy at a stress amplitude of 340 MPa, (c) fatigue strength of steel at 10,000 cycles.

Solution By reference to Fig. 13.5,

(a) The fatigue strength of the *Al* alloy at 10^7 cycles = 160 MPa
(b) The fatigue life of the aluminum alloy at a stress amplitude of 340 MPa = 10,000 cycles
(c) The fatigue strength of steel at 10,000 cycles = 480 MPa

Example 13.7 Calculating the Maximum Mean Stress by using Goodman's Law A rotating steel shaft is subjected to a stress range of 380 MPa. The tensile strength of the shaft steel is 500 MPa Calculate the maximum mean stress the steel shaft is able to withstand.

Solution $\sigma_r = 380$ MPa, $\sigma_{ut} = 500$ MPa, $\sigma_m = ?$.

By using Eqs. 13.3 and 13.6,

$$\sigma_a = \frac{\sigma_r}{2} = \frac{380}{2} = 190\,\text{MPa}$$

$$S_e = \frac{\sigma_{ut}}{2} = 0.5 \times 500 = 250\,\text{MPa}$$

By using Goodman's law (Eq. 13.7),

$$\frac{190}{250} + \frac{\sigma_m}{500} = 1$$

Mean stress $= \sigma_m = 0.24 \times 500 = 120\,\text{MPa}$

Example 13.8 Calculating the Stress Concentration Factor for Fatigue Loading. A steel notch-free bar exhibits a fatigue limit of 270 MPa. The same material exhibits a fatigue limit of 150 MPa to a notch effect. The static stress concentration factor for the bar is 5. Calculate the stress concentration factor for fatigue loading of the bar.

Solution By using Eq. 13.9,

Fatigue stress concentration factor $= K_f = \dfrac{Fatigue\ limit\ for\ notch-free\ specimen}{Fatigue\ limit\ for\ notched\ specimen} = \dfrac{270}{150} = 1.8$

Example 13.9 Calculating the Notch Sensitivity Factor for a Component By using the data in Example 13.8, calculate the notch sensitivity factor for the component.

Solution $K_s = 5, K_f = 1.8, q_n = ?$.

By using Eq. 13.8,

$$\text{Notch sensitivity factor} = q_n = \frac{K_f - 1}{K_s - 1} = \frac{1.8 - 1}{5 - 1} = \frac{0.8}{4} = 0.2$$

Example 13.10 Calculating the Permissible Number of Cycles for a Specified Fatigue Load Figure 13.9 shows the S-N curve for a hypothetical material in which σ_f denotes the failure stress. The fatigue behavior of the material can be described by the relation: $\log N = 12\left(1 - \dfrac{S}{\sigma_f}\right)$.

Calculate the permissible number of cycles when the load is $S = 0.8\ \sigma_f$.

Solution From Fig. 13.9, the material has been subjected to $n_1 = 10^4$ cycles at $S = 0.65\ \sigma_f$.

Thus the fatigue life at $S = 0.65\ \sigma_f$ can be calculated as follows:

$$\log N = 12\left(1 - \frac{S}{\sigma_f}\right) = 12\left(1 - \frac{0.65\sigma_f}{\sigma_f}\right) = 12(1 - 0.65) = 4.2$$

13.or$N_1 = 10^{4.2} = 1.6\ 10^4 = 16000$ cycles

It means that for $n_1 = 10^4$ cycles, $N_1 = 1.6 \times 10^4$ cycles. For $S = 0.8\ \sigma_f$, the fatigue life is:

$$\log N = 12\left(1 - \frac{0.8\sigma_f}{\sigma_f}\right) = 12(1 - 0.8) = 2.4$$
$$N_2 = 10^{2.4} = 251\,\text{cycles}$$

At stress $S = 0.8\,\sigma_f$, the number of applied cycles n_2 can be calculated by using Eq. 13.11,

Fig. 13.9 S-N curve for a hypothetical material

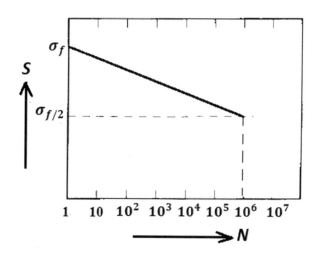

$$n_2 = N_2 \left(1 - \frac{n_1}{N_1}\right) = 251\left(1 - \frac{10000}{16000}\right) = 251(1 - 0.625) = 94 \, \text{cycles}$$

Hence, the permissible number of cycles at the stress $S = 0.8\sigma_f$ is 94.

Example 13.11 Calculating the Cumulative Damage to a Bridge for All Vehicles Traffic A large number of trains, trucks, and cars pass through a traffic bridge daily. The bridge has been in service for 2 years. The fatigue data for traffic on the bridge is given in Table 13.1.

In view of the un-expected heavy traffic on the bridge, a nearby second bridge has been recently constructed that can carry all train traffic. Calculate the cumulative damage to the first bridge for all traffic during the service period.

Solution

$$\text{Cycle ratio for daily damage due to train traffic} = \frac{n_1}{N_1} = \frac{22}{100,000} = 0.00022$$

$$\text{Cycle ratio for daily damage due to truck traffic} = \frac{n_2}{N_2} = \frac{100}{2 \times 10^6} = 0.00005$$

$$\text{Cycle ratio for daily damage due to car traffic} = \frac{n_3}{N_3} = \frac{5000}{10^{10}} = 0.0000005$$

Cumulative daily damage to the bridge =

$$\frac{n_1}{N_1} + \frac{n_2}{N_2} + \frac{n_3}{N_3} = 0.00022 + 0.00005 + 0.0000005 = 0.00027$$

The number of service days of the bridge $= 2 \, \text{years} \times 365 = 730$

The cumulative damage to the bridge in 2 years $= 730 \times 0.00027 = 0.197$

Example 13.12 Calculating the Fraction of Remaining Lifetime of Bridge By using the data in Example 13.10, calculate the fraction of remaining lifetime of the first bridge corresponding to trucks and cars traffic.

Solution By using Eq. 13.12,

The fraction of remaining lifetime of the bridge due to trucks and cars traffic = $1 - 0.197 = 0.803$.

Table 13.1 The fatigue data for the traffic bridge

#	Vehicle	Vehicle passing daily	Fatigue life, cycles
1.	Trains	22	10^5
2.	Trucks	100	2×10^6
3.	Cars	5000	10^{10}

Example 13.13 Calculating the Remaining Fatigue Lifetime of Bridge for Selected Traffic By using the data in Examples 13.11 and 13.12, calculate the remaining fatigue lifetime of the first bridge in days corresponding to the trucks and cars traffic only.

Solution Let number of days of the first bridge's remaining lifetime $= d$

$$\text{The remaining lifetime of the bridge due to trucks and cars traffic} = d \times \left(\frac{n_2}{N_2} + \frac{n_3}{N_3} \right)$$

$$= d \times (0.00005 + 0.0000005) = 5.05 \times 10^{-5} d$$

By using the data from Example 13.12,

$$5.05 \times 10^{-5} d = 0.803$$

13. or

$$d = \frac{0.803}{5.05 \times 10^{-5}} = 15900 \, (\text{days})$$

It means that by diverting the train traffic to the second bridge, the first bridge can be safely used in service for additional 15,900 days or around 43 years.

Example 13.14 Determining the Constants for Computing Fatigue Life of Cracked Structure A steel structure of semi-infinite width contains a 7.6-mm-deep edge crack that is oriented normal to the tensile stress. The structure is subject to fluctuating load with stress variation from 172 to 310 MPa. The plain strain fracture of the steel is $165\,\text{MPa}\sqrt{m}$. The fatigue crack growth rate obeys the following Paris law: $\dfrac{da}{dN} = 1.35 \times 10^{-10} (\Delta K)^{2.25}$.

where da/dN is in m/cycle. Determine the stress range $\Delta\sigma$, the geometric factor Y, crack size a_0 in m,, and the constants A and m for the Paris equation.

Solution
$$Y = 1.12 \, (\text{see Table 12.1}), a_0 = 7.6\,\text{mm} = 0.0076\,\text{m}, A = 1.35 \times 10^{-10}, m = 2.25$$
$$\Delta\sigma = \sigma_{\max} - \sigma_{\min} = 310 - 172 = 138\,\text{MPa}$$

Example 13.15 Calculating the Final/Critical Crack Size at Failure By using the data in Example 13.14, calculate the final (critical) crack size at failure.

Solution $\sigma_c = \sigma_{max} = 310$ MPa, $K_{IC} = 165\,\text{MPa}\sqrt{m}$, $a_f = ?$ Edge crack.

By using Eq. 12.19 for the critical (final) crack size,

$$a_f = a_c = \frac{1}{\pi}\left(\frac{K_{IC}}{Y\sigma_c}\right)^2 = \frac{1}{\pi}\left(\frac{165}{1.12 \times 310}\right)^2 = 0.318 \times 0.226 = 0.0718\,\mathrm{m}$$

The final crack size at failure = $a_f = 0.0718$ m.

Example 13.16 Performing Integration to Derive a Formula to Compute Fatigue Life By using calculus technique, perform definite integration of Eqs. 13.17a, and 13.17b to derive a formula for computing the fatigue life of cracked component.

Solution By re-arranging the terms in Eqs. 13.17a and 13.17b,

$$\int_0^{N_f} dN = \int_{a_0}^{a_f} \frac{da}{A\left(Y\left(\Delta\sigma\right)\sqrt{\pi a}\right)^m} = \int_{a_0}^{a_f} \frac{da}{AY^m\left(\Delta\sigma\right)^m\left(\pi a\right)^{\frac{m}{2}}} = \frac{1}{AY^m\left(\Delta\sigma\right)^m\pi^{\frac{m}{2}}}\int_{a_0}^{a_f} a^{-m/2}\,da$$

$$\left(N_f - 0\right) = \frac{1}{AY^m\left(\Delta\sigma\right)^m\pi^{\frac{m}{2}}}\left(\frac{2a_f^{\frac{2-m}{2}}}{2-m} - \frac{2a_0^{\frac{2-m}{2}}}{2-m}\right)$$

$$N_f = \frac{2}{AY^m\left(\Delta\sigma\right)^m\pi^{\frac{m}{2}}}\left(\frac{a_f^{\frac{2-m}{2}}}{2-m} - \frac{a_0^{\frac{2-m}{2}}}{2-m}\right) \tag{13.18}$$

Example 13.17 Calculating the Fatigue Life of a Cracked Component By using the data in *Examples 13.14 – 13.15* and by using the formula derived in *Example 13.16*, calculate the fatigue life of the cracked component.

Solution $Y = 1.12$, $a_0 = 00076$ m, $a_f = 0.0718$ m, $\Delta\sigma = 138$ MPa, $A = 1.35 \times 10^{-10}$, $m = 2.25$, $K_{IC} = 165$ MPa$\sqrt{m}$,

$$N_f = \frac{2}{AY^m\left(\Delta\sigma\right)^m\pi^{\frac{m}{2}}}\left(\frac{a_f^{\frac{2-m}{2}}}{2-m} - \frac{a_0^{\frac{2-m}{2}}}{2-m}\right)$$

$$N_f = \frac{2}{1.35 \times 10^{-10} \times 1.12^{2.25}\left(138\right)^{2.25}\pi^{1.125}}\left(\frac{0.0718^{-0.125}}{-0.25} - \frac{0.0076^{-0.125}}{-0.25}\right)$$

$$N_f = \frac{2}{412058 \times 10^{-10}}\left(\frac{1.3899}{-0.25} - \frac{1.84}{-0.25}\right) = 48500\left(7.36 - 5.56\right) = 87,300\,\mathrm{cycles}$$

Example 13.18 Calculating the Stress Range for a Desired Fatigue Life A metal plate has a crack with $a_0 = 0.25$ mm. The plate is to be subjected to alternating stress cycle. The critical crack size is 5 mm, and the geometric factor is 2.0. The desired fatigue life is 320,000 cycles. The Paris equation is: $\dfrac{da}{dN} = 5 \times 10^{-15} (\Delta K)^4$. Calculate the stress range.

Solution $a_0 = 0.00025$ m, $a_f = 0.005$ m, $Y = 2$, $A = 5 \times 10^{-15}$, $m = 4$, $N_f = 320,000$ cycles, $\Delta\sigma = ?$.

By using Eq. 13.18,

$$N_f = \frac{2}{A Y^m (\Delta\sigma)^m \pi^{\frac{m}{2}}} \left(\frac{a_f^{\frac{2-m}{2}}}{2-m} - \frac{a_0^{\frac{2-m}{2}}}{2-m} \right)$$

$$320,000 = \frac{2}{5 \times 10^{-15} \times 2^4 (\Delta\sigma)^4 \pi^2} \left(\frac{0.005^{-1}}{-2} - \frac{0.00025^{-1}}{-2} \right)$$

The stress range $= \Delta\sigma = 350 \, \text{MPa}$

Questions and Problems

13.1. Explain fatigue failure mentioning the three stages. List six components that fail by fatigue.

13.2. Draw labelled diagrams showing the three types of stress cycles. Give an example for each type of stress cycle. What is the value of mean stress in alternating stress cycle?

13.3. Define the terms: (a) fatigue strength, (b) fatigue life, (c) fatigue limit, and (d) HCF.

13.4. Draw Goodman's diagram for alternating stresses.

13.5. What are the techniques for designing against fatigue failure? Explain any technique.

13.6. (a) Draw the diagram showing the variation of fatigue crack growth rate on σ and a. (b) Derive Eqs. 13.17a and 13.17b and give the meaning of each term in the eq.

13.7. The alternating stresses acting on a component are ±85 MPa. Calculate the: (a) mean stress, (b) stress range, (c) stress amplitude, (d) stress ratio, and (e) amplitude ratio.

13.8. By reference to Fig. 13.9, calculate the permissible number of cycles at the stress $S = 0.9 \, \sigma_f$ for the material.

13.9. A rotating steel shaft is subjected to a stress range of 300 MPa. The tensile strength of the steel is 480 MPa Calculate the maximum mean stress the steel shaft is able to withstand.

Table 13.2 The fatigue data for the traffic through the bridge

#	Vehicle	Vehicle passing daily	Fatigue life, cycles
1.	Trucks	50	10^4
2.	Cars	130	7×10^6
3.	Motor-bikes	8000	10^{12}

13.10. A metal plate of semi-infinite width has a 0.8-mm-long edge crack. The plate is to be subjected to alternating stress cycle with a mean stress of 30 MPa. The critical crack size is 7 mm. The desired fatigue life is 280,000 cycles. Calculate the maximum tensile stress to yield the prescribed fatigue life. The Paris equation is: $\dfrac{da}{dN} = 3 \times 10^{-12} \left(\Delta K \right)^3$.

13.11. A steel notch-free bar exhibits a fatigue limit of 320 MPa. The same material exhibits a fatigue limit of 200 MPa to a notch effect. The static stress concentration factor for the bar is 6. Calculate the notch sensitivity factor for the component.

13.12. A large number of trucks, cars, and motor-bikes pass through a traffic bridge daily. The bridge has been in service for 3 years. Table 13.2 gives the fatigue data for the traffic.

In view of the un-expected heavy traffic on the bridge, an alternative road has been recently constructed through which all truck traffic can pass. Calculate the remaining fatigue lifetime of the bridge in days corresponding to the car and motor-bike traffic only.

13.13. (MCQs). Encircle the correct answers for each of the following questions.

(a) Which law deals with the fatigue crack growth rate?
   (i) Miner's law, (ii) Paris law, (iii) Goodman's law, (iv) Griffith law.

(b) Which law deals with relationship to compute the mean stress?
   (i) Miner's law, (ii) Paris law, (iii) Goodman's law, (iv) Griffith law.

(c) Which law deals with the cumulative damage to a component due to cyclic stresses?
   (i) Miner's law, (ii) Paris law, (iii) Goodman's law, (iv) Griffith law.

(d) Which type of stress cycle exists in automobile shock absorbers?
   (i) alternating, (ii) repeating, (iii) fluctuating, (iv) uniform.

(e) Which type of stress cycle exists in a rotating shaft?
   (i) alternating, (ii) repeating, (iii) fluctuating, (iv) uniform.

(f) Which type of stress cycle exists in a cantilever beam?
   (i) alternating, (ii) repeating, (iii) fluctuating, (iv) uniform.

References

Huda Z, Bulpett R (2012) Materials science and Design for Engineers. Trans Tech Publications, Switzerland

Huda Z, Ajani MH, Ahmed MS (2020) Fatigue behaviors of two notched cutting-tool materials: high speed steel and cemented carbide. Materials Testing 62(3):265–270

Rajabipour A, Melchers RE (2015) Application of Paris' law for estimation of hydrogen- assisted fatigue crack growth. Int J Fatigue 80:357–363

Schijve J (2009) Fatigue of structures and materials. Springer Publishing, Berlin, Germany

Chapter 14
Creep Behavior of Materials

14.1 Creep Deformation and Failure

Creep refers to a time-dependent deformation at a constant stress, usually at an elevated temperature. An excessive creep deformation may result in failure – *creep failure*. The term *stress rupture* refers to creep failure by fracture. At temperatures exceeding 0.45 of the melting temperature (in Kelvin), materials subjected to constant stress will undergo creep failure (Lokoshchenko 2018). Materials and components may fail by creep at stresses far below their yield strengths. Examples of machine components that experience creep failure include: gas-turbine engines' components, nuclear reactors, oven components, and boilers' superheater tubes (see Fig. 14.1). Creep also occurs at ambient temperatures; for example, creep occurs in pre-stressed concrete beams, which are held in compression by steel rods.

The mechanism of creep involves dislocation climb; which is a thermally activated diffusive process. Figure 14.2 shows the climb of an edge dislocation; here the dislocation line climbs by an interatomic distance when a job travels along its length by absorbing vacancies. Creep failure usually involves three stages of deformation: (a) primary creep, (b) secondary creep, and (c) tertiary creep leading to failure (see Fig. 14.3).

14.2 Creep Testing and Creep Curve

Creep testing involves the application of a constant load to a specimen at a high temperature (see Fig. 14.4). A creep test is used to determine the amount of deformation a material experiences over time while under a continuous tensile or compressive load at a constant (elevated) temperature. Thus the loaded material undergoes progressive rate of deformation. Figure 14.4 shows basic elements of a

Z. Huda, *Mechanical Behavior of Materials*, Mechanical Engineering Series, https://doi.org/10.1007/978-3-030-84927-6_14

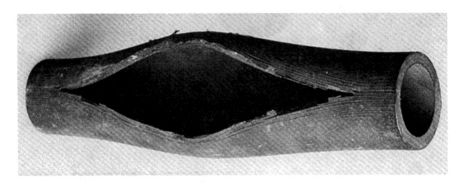

Fig. 14.1 Creep failure in a boiler's superheater tube due to long-time overheat. (*Courtesy –* Reprinted by Permission of *Babcock & Wilcox Company*, Akron, USA)

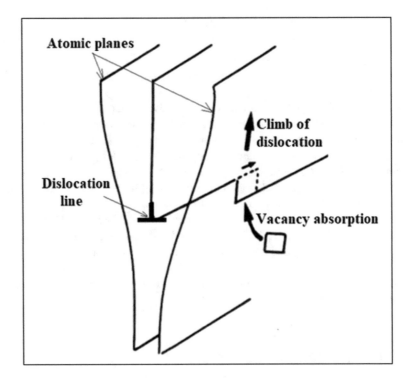

Fig. 14.2 Creep by dislocation climb mechanism

creep testing machine with variable lever arms to ensure constant stress on the specimen; it must be noted that the length l_2 decreases as the length of the specimen increases. The corresponding amount of strain, in the specimen, is recorded over the period of time, as shown by a typical creep curve in Fig. 14.5.

It is evident in the creep curve (Fig. 14.5) that the creep deformation begins with an initial elastic strain followed by three stages in creep: (I) primary creep, (II)

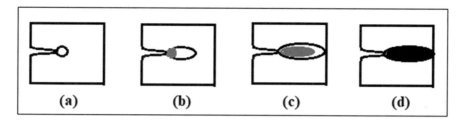

Fig. 14.3 Creep crack growth stages; (a) primary creep, (b,c) secondary creep, and (c) tertiary creep leading to failure

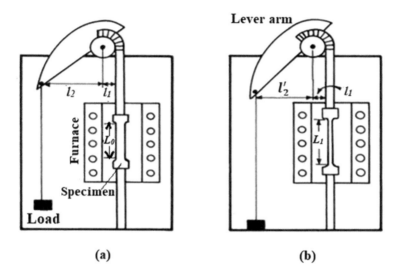

Fig. 14.4 Creep testing principle; (a) dimensions before the application of load, (b) the dimensions after application of the load

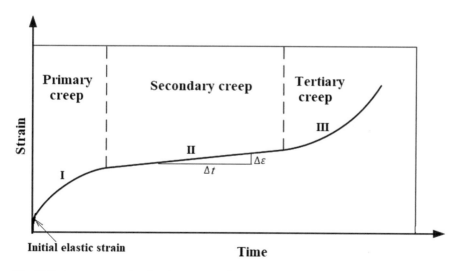

Fig. 14.5 A creep curve showing three stages of creep

secondary creep, and (III) tertiary creep (see also Fig. 14.3). The stage II (secondary creep) shows a steady-state creep behavior; here, the rate of creep is determined by the slope of the curve, as follows:

$$\text{Creep rate} = \dot{\varepsilon} = \frac{\Delta\varepsilon}{\Delta t} \tag{14.1}$$

where $\Delta\varepsilon$ is the strain increment over the time period Δt, h^{-1} (see Examples 14.1 and 14.2).

14.3   Factors Controlling Creep Rate

The rate of creep of a material is strongly dependent on stress and temperature. Besides stress and temperature, the creep rate also depends on the grain size at a temperature above 0.5 of the melting temperature (in Kelvin). It has been reported that the steady-state creep rate decreases with increasing grain size up to some optimum grain size (~100 μm); for grain size above the optimum value, the creep rate is independent of grain size (Malakondaiah and Rao 1985).

The rate of creep varies directly as the applied stress. The steady-state creep strain rate $(\dot{\varepsilon})$ is related to the applied stress (σ) at constant temperature by:

$$\dot{\varepsilon} = C_1 \sigma^n \tag{14.2}$$

where C_1 and n are the material's constants (see Examples 14.3, 14.4, and 14.5).

The creep rate is significantly increased by an increase in temperature. The steady-state creep rates vary exponentially with temperature according to the Arrhenius-type expression:

$$\dot{\varepsilon} = C_2 \sigma^n \exp\left(-\frac{Q_c}{RT}\right) \tag{14.3}$$

where $\dot{\varepsilon}$ is the creep rate, s^{-1}; T is the temperature, K; C_2, n and the stress σ are constants; R = 8.3145 J($mol \cdot K$)$^{-1}$; and Q_c is the activation energy for creep, J $\cdot mol^{-1}$. The activation energy for creep is the minimum energy required to cause creep. By equating the term $C_2\sigma^n$ to another constant A, Eq. 14.3 is simplified as:

$$\dot{\varepsilon} = A\exp\left(-\frac{Q_c}{RT}\right) \tag{14.4}$$

By taking the logarithms of both sides, Eq. 14.4 takes the form:

$$\log\dot{\varepsilon} = \log A - \frac{Q_c}{RT} \tag{14.5}$$

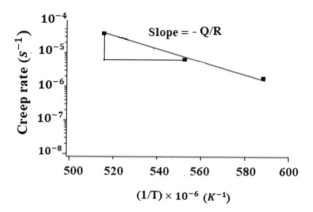

Fig. 14.6 Log-linear plot of minimum creep rate versus reciprocal of temperature for a polycrystalline oxide (stress = 80 MPa)

Thus, the activation energy for creep, Q_c, can be determined experimentally, by plotting the logarithm of creep rate against the reciprocal of temperature (see Fig. 14.6).

The significance of Eq. 14.5 and Fig. 14.6 is illustrated in Examples 14.6, 14.7, 14.8, and 14.9.

14.4 Larson-Miller Parameter (*LMP*)

The Larson-Miller parameter (*LMP*) is a useful tool for calculating the stress-rupture time or life of a component at various temperatures for a given material. The LMP describes the equivalence of time at a temperature for an alloy for stress rupture. It is mathematically defined as:

$$LMP = \frac{T}{1000}(C + \log t) \qquad (14.6)$$

where *LMP* is the Larson-Miller parameter; *T* is the temperature in Rankine (°R); *t* is the stress-rupture time or creep-life, h; and C is a material's constant (usually C = 20). Accordingly, Eq. 14.6 takes the form:

$$LMP = T(20 + \log t) \times 10^{-3} \qquad (14.7)$$

Fig. 14.7 illustrates the LMPs for some titanium alloys (see Examples 14.10 and 14.11). Experiments have shown that the creep lives for most materials lie in the range of 100–100,000 h.

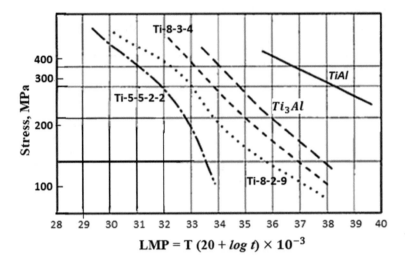

Fig. 14.7 The plots showing Larson-Miller parameters (LMP) at various stresses for some titanium alloys

14.5   Creep-Limited Alloy Design

In general, ceramics have excellent creep resistance. The alloy design techniques that can be applied to limit creep are listed as follows: (a) select high melting temperature elements, (b) add solutes that raise the modulus of elasticity or lower the stacking fault energy, (c) precipitation harden to produce fine, dispersed second-phase particles to obstruct dislocations, (d) increase grain size by normal grain-growth annealing heat treatment, (e) dispersion harden the alloy to produce nano-sized dispersed phase particles. Nickel-base superalloys are designed by following the above-listed rules; these alloys are extensively used in hot-sections of gas-turbine engines and nuclear reactors (see Chap. 5). In particular, dispersion strengthened superalloys (*e.g.* MA-753, MA-6000, etc.) possess excellent creep resistance; they contain nano-sized disperoids (*e.g.* yttrium oxide) with inter-particle spacing <500 nm in their microstructures Creep-resistant single-crystal (SC)/columnar grained (CG) superalloy turbine-blades are manufactured by directional solidification (DS) casting process (Huda 2020, 1995, 2007).

14.6   Calculations – *Worked Examples*

Example 14.1 Calculating the Strains for various Time Durations during Creep Test A tensile specimen, with the original gage length of 50 mm, was creep tested at 800 °C. The length increased to 52 mm in 20 h. Another reading of the creep data showed the new length of 56 mm in 50 h. Calculate the strains for the two time durations.

Solution For $t_1 = 20$ h duration, $\Delta l_1 = l - l_0 = 52 - 50 = 2$ mm; so the strain is:

$$\varepsilon_1 = \frac{\Delta l_1}{l_0} = \frac{2}{50} = 0.04$$

For $t_2 = 50$ h duration, $\Delta l_2 = l - l_0 = 56 - 50 = 6$ mm; so the new strain is:

$$\varepsilon_2 = \frac{\Delta l_2}{l_0} = \frac{6}{50} = 0.12$$

Example 14.2 Calculating the Creep Rate by using Creep Data By using the data in Example 14.1, calculate the creep rate.

Solution For $t_1 = 20$ h, $\varepsilon_1 = 0.04$; for $t_2 = 50$ h, $\varepsilon_2 = 0.12$

$$\Delta\varepsilon = \varepsilon_2 - \varepsilon_1 = 0.12 - 0.04 = 0.08; \Delta t = t_2 - t_1 = 50 - 20 = 30\text{h}$$

By using Eq. 14.1,

$$\text{Creep rate} = \dot{\varepsilon} = \frac{\Delta\varepsilon}{\Delta t} = \frac{0.08}{30} = 2.67 \times 10^{-3}\,h^{-1}$$

Example 14.3 Calculating the Constants n and C_1 in Eq. 14.2 A stainless steel specimen was creep tested at 520 °C at a stress of 420 MPa for 400 hours; thereby producing a strain of 0.14. The same specimen was again creep tested at the same temperature at a stress of 310 MPa for 1000 hours that produced a strain of 0.08. Assuming steady-state creep, calculate the constants n and C_1 in Eq. 14.2.

Solution. At a stress of 420 MPa,

$$\dot{\varepsilon}_1 = \frac{\Delta\varepsilon}{\Delta t} = \frac{0.14}{400} = 350 \times 10^{-6}\,h^{-1} \text{ (see Eq. 14.1).}$$

At a stress of 310 MPa,

$$\dot{\varepsilon}_2 = \frac{\Delta\varepsilon}{\Delta t} = \frac{0.08}{1000} = 80 \times 10^{-6}\,h^{-1} \text{ (see Eq. 14.1).}$$

By using Eq. 14.2 $\left(\dot{\varepsilon} = C_1 \sigma^n\right)$, we can write:

$$350 \times 10^{-6} = C_1 \times 420^n \qquad \text{(14.8a)}$$

$$80 \times 10^{-6} = C_1 \times 310^n \qquad \text{(14.8b)}$$

By simultaneously solving Eqs. 14.8a and 14.8b, we obtain:

$$\left(\frac{420}{310}\right)^n = \frac{350}{80} = 4.37$$

or $n = 4.92$

By substituting the value of n in Eq. 14.8a, we obtain $C_1 = 4.32 \times 10^{-17}$.

Example 14.4 Calculating the Creep Rate at a Specified Stress for a Material By using the data in Example 14.3, calculate the creep rate in the material at a stress of 100 MPa.

Solution $n = 4.92$, $C_1 = 4.32 \times 10^{-17}$, $\sigma = 100$ MPa.

By using Eq. 14.2,

$$\dot{\varepsilon} = C_1 \sigma^n = 4.32 \times 10^{-17} \times 100^{4.92} = 4.32 \times 10^{-17} 6.91 \times 10^9 = 3 \times 10^{-7} h^{-1}$$

Example 14.5 Calculating the Time to Creep for a Given Stress By using the data in Example 14.3, calculate the time to produce 0.11% strain in a link bar of the same material when stressed to 80 MPa at the same temperature (520 °C).

Solution For a stress of 80 MPa and strain change of 0.11%, the creep rate is:

$$\text{Creep rate} = C_1 \sigma^n = 4.32 \times 10^{-17} (80)^{4.92} = 9.97 \times 10^9 \times 10^{-17} = 9.97 \times 10^{-8} h^{-1}$$

Now, creep strain rate $= \dfrac{\Delta \varepsilon}{\Delta t} = 9.97 \times 10^{-8}$ (see Eq. 14.1)

or $\Delta t = \dfrac{\Delta \varepsilon}{creep\,rate} = \dfrac{0.0011}{9.97 \times 10^{-8}} = 11{,}030h$

Time to creep = 11,030 hours.

Example 14.6 Calculating the Activation Energy for Creep by Analytical Method A creep test on a material at a stress of 150 MPa results in the steady-state creep rate of 3.0×10^{-4} s^{-1} at 870 K; and the creep rate is 5.4×10^{-8} s^{-1} at a temperature of 770 K. Calculate the activation energy for creep for the temperature range for the material.

Solution By using Eq. 14.5 for T = 870 K and $\dot{\varepsilon} = 3.0 \times 10^{-4}\,s^{-1}$, we can write:

$$\log \dot{\varepsilon} = \log A - \frac{Q_c}{RT}$$

$$\log 3.0 \times 10^{-4} = \log A - \frac{Q_c}{8.31 \times 870}$$

or

$$-3.52 = \log A - \frac{Q_c}{7230} \qquad (14.9a)$$

By using Eq. 14.5 for T = 770 K and $\dot{\varepsilon} = 5.4 \times 10^{-8}\,s^{-1}$,

$$\log 5.4 \times 10^{-8} = \log A - \frac{Q_c}{8.31 \times 770}$$

or

$$-7.27 = \log A - \frac{Q_c}{6399} \qquad (14.9b)$$

By subtracting Eq. 14.9b from Eq. 14.9a, we obtain:

$$-3.52 + 7.27 = \frac{Q_c}{6399} - \frac{Q_c}{7230}$$

$$Q_c \left(0.000156.3 - 0.000138.3 \right) = 3.75$$

$$Q_c = 208,300\,J\,/\,mol = 208.3\,kJ\,/\,mol$$

The activation energy for creep in the temperature range of 770–870 K = 208.3 kJ/mol.

Example 14.7 Determination of the Activation Energy for Creep by Graphical Method By reference to the graphical plot in Fig. 14.6, determine the activation energy for creep for the temperature range of 1808–1920 K for the polycrystalline oxide.

Solution The reciprocals of given temperatures are: $1/T_1 = (520 \times 10^{-6})K^{-1}$, $1/T_2 = (553 \times 10^{-6})K^{-1}$.

From the graphical plot (Fig. 14.6),

$$\text{Slope} = -\frac{Q_c}{R} = \frac{\left[\log\left(5.8\times10^{-5}\right)-\left[\log\left(2.3\times10^{-6}\right)\right]\right]}{\left(553\times10^{-6}\right)-\left(520\times10^{-6}\right)} = 42,000K^{-1}$$

$$Q_c = 42,000\times R = 42,000\times8.314 = 349,188 = 349.2 \text{ kJ}mol^{-1}$$

Example 14.8 Calculating the Maximum Working Temperature for Limiting Creep By using the data in *Example 14.6*, calculate the maximum working temperature for the material if the limiting creep rate is 3×10^{-6} s^{-1}.

Solution $\dot{\varepsilon} = 3\times10^{-6}s^{-1}$, $Q_c = 208, 300$ J/mol., $T =$?

By using Eq. 14.9a,

$$-3.52 = \log A - \frac{Q_c}{7230}$$

$$-3.52 = \log A - \frac{208300}{7230}$$

$$\log A = 25.28$$

By using Eq. 14.5,

$$\log \dot{\varepsilon} = \log A - \frac{Q_c}{RT}$$

$$\log 3\times10^{-6} = 25.28 - \frac{208300}{8.31T}$$

$$T = 813.8K$$

The maximum working temperature $= 813.8\,K = 540.8°\,C$

Example 14.9 Calculating the Creep Life at a Specified Stress and Temperature By using the data in Example 14.8, calculate the creep life of the material at a stress of 150 MPa at the maximum working temperature of 541 °C if the total creep strain is up to 0.2.

Solution At   the   maximum   working   temperature   (541   °C), the creep strain rate $= \dot{\varepsilon} = 3\times10^{-6}s^{-1}$

$$\Delta\varepsilon = 0.2, \Delta t = ?$$

By using the re-arranged form of Eq. 14.1,

$$\Delta t = \frac{\Delta \varepsilon}{creep\,rate} = \frac{0.2}{3 \times 10^{-6}} = 6670s = 18.5h$$

The creep life of the material = 18 hours.

Example 14.10 Calculating the Stress-Rupture Time or Creep Life for a Material By reference to the Larson-Miller parameter data in Fig. 14.7, calculate the time to rupture the *TiAl* alloy component that is subjected to a stress of 300 MPa at a temperature of 600 °C.

Solution By reference to Fig. 14.6 for *TiAl* alloy at 300 MPa stress, the value of LMP is 37.8.

By using Eq. 14.7,

$$37.8 = T(20 + \log t) \times 10^{-3}$$

At 600 °C or T = (600 × 1.8) + 491.67 = 1571.67°R

$$37.8 = 1571.67(20 + \log t) \times 10^{-3}$$

$$\log t = 4$$

$$t = 10^4 h = 10{,}000h$$

The stress rupture time of the component = $10{,}000\,h = 10{,}000 \times 0.000114155$
= 1.14 year

Example 14.11 Determining the Design Stress by using L-M Parameter By reference to Fig. 14.7, what design stress do you recommend for Ti-8-2-9 alloy component exposed to a temperature of 500 °C, if the component's life is desired to be 2 years?

Solution T = 500 °C = 1391.67 °R, t = 2 years = 8760 × 2 = 17520 h.

By using Eq. 14.7,

$$LMP = T(20 + \log t) \times 10^{-3} = 1391.67(20 + \log 17520) \times 10^{-3} = 1.39 \times 24.24 = 33.7$$

By reference to Fig. 14.6, LMP = 33.7 for Ti-8-2-9 alloy corresponds to 220 MPa. Hence, the safe design stress should be 200 MPa.

Questions and Problems

14.1. What is the difference between creep failure and stress rupture?

14.2. Give four examples of components that frequently fail by creep.

14.3. What is the effect of grain size of a material on the creep rate?

14.4. Draw a diagram showing the creep mechanism by dislocation climb.

14.5. Explain the three stages of creep with the aid of sketch of creep curve.

14.6. A stainless steel specimen was creep tested at 470 °C at a stress of 380 MPa for 360 hours; thereby producing a strain of 0.12. The same specimen was again creep tested at the same temperature at a stress of 270 MPa for 600 hours that produced a strain of 0.03. Calculate the creep rate in the material at a stress of 140 MPa.

14.7. The constants for creep behavior of a material are: $n = 4.13$, and $C_1 = 3.95 \times 10^{-18}$ at a temperature of 500 °C. Calculate the time to produce 0.28% strain in a link bar of the same material when stressed to 120 MPa at the same temperature (500 °C).

14.8. A creep test on a material at a stress of 200 MPa results in the steady-state creep rate of $2.0 \times 10^{-3} s^{-1}$ at 850 K; and the creep rate is $5.4 \times 10^{-6} s^{-1}$ at a temperature of 730 K. Calculate the activation energy for creep for the temperature range for the material.

14.9. By reference to the Larson-Miller parameter data in Fig. 14.7, calculate the time to rupture the Ti_3Al alloy component that is subjected to a stress of 200 MPa at 450 °C.

14.10. By using the data in Problem 14.8, calculate the maximum working temperature for the material.if the limiting creep rate is $4.5 \times 10^{-5} s^{-1}$.

14.11. By reference to Fig. 14.6, what design stress do you recommend for Ti-8-3-4 alloy exposed to a temperature of 550 °C, if the component's life is desired to be 3 years?

14.12. (MCQs). Encircle the correct answers for the following questions.

(a) Which stage of creep enables us to calculate the creep rate?

(i) Primary creep, (ii) secondary creep, (iii) tertiary creep.

(b) Which term corresponds to fracture of a component?

(i) stress rupture, (ii) creep, (iii) creep failure, (iv) deformation.

(c) Which alloy has the best creep resistance?

(i) brass, (ii) bronze, (iii) aluminum alloy, (iv) superalloy.

(d) Which scale of temperature is used in calculating the Larson-Miller parameter?

(i) °C, (ii) °F, (iii) °R, (iv) Kelvin.

References

Huda Z (2020) Metallurgy for physicists and engineers. CRC Press, Boca Raton, FL, USA

Huda Z (2007) Development of heat treatment process for P/M superalloy turbine blades. Materials Design 28(5):1664–1667

Huda Z (1995) Development of design principles for a creep-limited alloy for turbine blades. J Mater Eng Perform 4(1):48–53

Lokoshchenko AM (2018) Creep and long-term strength of metals (1st edition). CRC Press, Boca Raton, FL

Malakondaiah G, Rao PR (1985) Effect of grain size, grain shape and subgrain size on high-temperature creep behavior. Defense Sci J 35(2):201–217

Chapter 15
Mechanical Behavior of Composite Materials

15.1 Composite Materials, Classification, and Applications

In Chap. 5, we learnt that a composite material is a combination of two materials with different properties. Based on the matrix material, there are three types of composites: (1) metal matrix composites (*MMC*), (2) polymer matrix composites (*PMC*), and (3) ceramic matrix composites. Based on the reinforcing material structure, there are three types of composite: (a) particulate composite, (b) fiber reinforced composite, and (c) laminate composite (see Fig. 15.1).

The three types of composites, based on reinforcing material structures, are illustrated in Fig. 15.2. The different types of composite material have been explained in Chap. 5. In fibrous composites, the orientation of fibers may be in one of the following ways: (a) unidirectional continuous fibers, (b) unidirectional discontinuous fibers, and (c) random discontinuous fibers (see Fig. 15.3). In addition to the various types of composite shown in Fig. 15.1, there are hybrid composites; which gave a combination of two or more reinforcement materials (see Fig. 15.4).

Some examples of composites include: wood, concrete, and advanced composites (e.g. graphite/epoxy, boron/aluminum composite, etc.). In wood, the lignin matrix is reinforced with cellulose fibers. Concrete is a particulate composites comprising of sand and cement. Advanced composites find wide applications in aerospace structures and defense industry. In particular, carbon fiber reinforced polymer (CFRP) composites, possessing the best specific modulus, are being used in aircrafts as well as in defense vehicles (see Table 15.1).

The specific modulus is defined as the ratio of the Young's modulus to the density. It is evident in Table 15.1 that the specific modulus of carbon fiber epoxy composite is 43.75 GPa/g·cm^{-3}; which is 3 times that of steel; which in turn justifies the wide aerospace applications of CFRP composites. Besides CFRP, glass fiber reinforced plastic (GFRP) and glass reinforced aluminum laminate (GLARE) composites are being used in both military and commercial aircrafts (see Fig. 15.5).

© The Author(s), under exclusive license to Springer Nature Switzerland AG 2022
Z. Huda, *Mechanical Behavior of Materials*, Mechanical Engineering Series,
https://doi.org/10.1007/978-3-030-84927-6_15

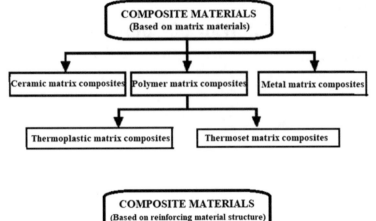

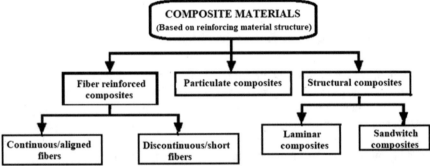

Fig. 15.1 Classification chart of composite materials

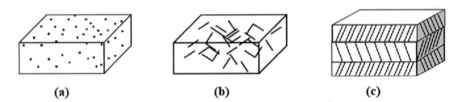

Fig. 15.2 Three types of composites; (**a**) particulate composite, (**b**) fibrous composite, and (**c**) laminate composite

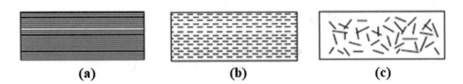

Fig. 15.3 Three types of fiber orientations in fibrous composites: (**a**) unidirectional continuous fibers, (**b**) unidirectional discontinuous fibers, and (**c**) random discontinuous fibers

Fig. 15.4 A hybrid
composite – glass
reinforced aluminum
laminate (GLARE)

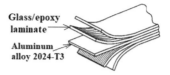

Glass/epoxy
laminate

Aluminum
alloy 2024-T3

Table 15.1 Some physical/mechanical properties of carbon fiber-epoxy composite

Property	Density	Young's modulus	Shear modulus	Shear strength	Compressive Strength	Coefficient of thermal expansion
Value	1.6 g cm⁻³	70 GPa	5 GPa	90 MPa	570 MPa	2.1×10^{-6}

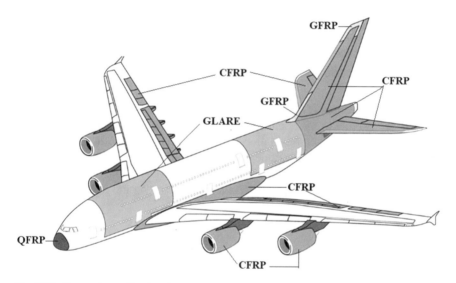

Fig. 15.5 Composite applications in commercial aircraft (QFRP = quartz fiber reinforced plastic composite)

15.2   Mechanical Behavior of Fibrous Composites

15.2.1   General Mechanical Behavior of Fibrous Composites

In order to effectively strengthen a fibrous composite, it is necessary that the fiber has an optimum or critical length to diameter ratio in the range of 30–140; which depends on the tensile strength of the fiber and the fiber-matrix (adhesive) bond strength. The critical fiber length to diameter ratio can be calculated by:

$$\frac{l_c}{d} = \frac{\sigma_f}{2\tau_c} \tag{15.1}$$

where l_c is the critical fiber length, mm; d is the diameter of the fiber, mm; σ_f is the tensile strength of the fiber, MPa, and τ_c is the adhesive bond strength, MPa (see Example 15.1).

In fibrous composites, the reinforcement material provides strength, stiffness, and impact strength, while the matrix material provides ductility, and resistances to fire and corrosion. In particular, continuous fiber-reinforced composite tubes possess excellent crushing behavior against vehicle accidental damages (Farley and Jones 1992). The stiffness and strength of a fibrous composite with unidirectional continuous fibers is strongly dependent on the orientation of fibers with respect to applied load (Gay 2015). The mechanical behavior of a fibrous composite can be measured by: (a) stiffness and strength under longitudinal loading, (b) stiffness under transverse loading, (c) Poisson's ratio, and (d) shear modulus.

15.2.2 Behavior of Unidirectional Continuous Fiber Composite under Longitudinal Loading

Consider a unidirectional continuous fiber composite under longitudinal loading *i.e.* all fibers are parallel to the applied stress (Fig. 15.6). In longitudinal loading, the stress on the composite causes uniform strain in all the composite layers *i.e.* the strain in matrix and fiber is the same (iso-strain condition) (Kaw 2006).

It is evident in Fig. 15.6 that the force applied to the composite (F or F_c) must be balanced by the equal and opposite tensile force acting on the fibers (F_f) and the force on matrix (F_m) *i.e.*

$$F = F_c = F_f + F_m \tag{15.2}$$

Fig. 15.6 A unidirectional continuous fiber composite under longitudinal loading

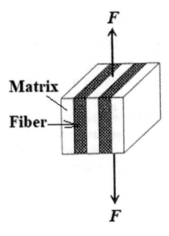

The fraction of force borne by the fibers phase can be calculated by analyzing Fig. 15.6.

Since the force in each phase is the stress (on the phase) times the area, Eq. 15.2 becomes:

$$\sigma_c A_c = \sigma_f A_f + \sigma_m A_m \tag{15.3}$$

By dividing both sides of Eq. 15.3 by cross-sectional area of the entire composite, we get:

$$\sigma_c = \sigma_f V_f + \sigma_m V_m \tag{15.4}$$

V_f is the volume fraction of the fiber phase, and V_m is the volume fraction of the matrix phase.

By combining Eq. 3.7 ($\sigma = E\ \varepsilon_e$) and Eq. 15.4, we obtain:

$$E_c \varepsilon_c = E_f \varepsilon_f V_f + E_m \varepsilon_m V_m \tag{15.5}$$

For compatibility, the strains in fibers, matrix, and composites are same *i.e.* $\varepsilon_c = \varepsilon_f = \varepsilon_m$ (as confirmed in Example 15.5). This is called iso-strain condition; so:

$$E_{cl} = E_f V_f + E_m V_m \tag{15.6}$$

where E_{cl} is the Young's modulus of the composite under longitudinal loading; E_f is the modulus of the fiber along its length; and E_m is the modulus of the matrix.

By combining Eqs. 15.2, 15.3, 15.4, 15.5, and 15.6, we obtain:

$$\frac{F_f}{F_m} = \frac{E_f V_f}{E_m V_m} \tag{15.7}$$

The significance of Eqs. 15.2, 15.3, 15.4, 15.5, 15.6, and 15.7 is illustrated in Examples 15.2, 15.3, 15.4, 15.5, and 15.6.

Similarly, the tensile strength of the entire composite under longitudinal loading is calculated by:

$$\sigma_{cl} = \sigma_f V_f + \sigma_m V_m \tag{15.8}$$

where σ_{cl} is the tensile strength of the entire composite under longitudinal loading; σ_f is the tensile strength of the fiber along the length of the fiber; and σ_m is the tensile strength of the matrix material (see Example 15.7).

15.2.3   Stiffness of Unidirectional Continuous Fiber Composite under Transverse Loading

Now we consider the unidirectional continuous fiber composite under transverse loading *i.e.* all the fibers are oriented perpendicular to the applied stress (see Fig. 15.7). Under transverse loading, the matrix and the fibers are under the same stress *i.e.* $\sigma_c = \sigma_f = \sigma_m$ (*iso-stress condition*).

The deformation (elongation) in the entire composite (δ_c) is the sum of deformations in the fibers phase (δ_f) and the matrix phase (δ_m) *i.e.*

$$\delta_c = \delta_m + \delta_f \tag{15.9}$$

Since $\delta = \varepsilon \cdot l$, Eq. 15.9 takes the form:

$$\varepsilon_c = \varepsilon_m V_m + \varepsilon_f V_f \tag{15.10}$$

By using Eq. 3.7 $\left(\varepsilon_e = \dfrac{\sigma}{E} \right)$, we obtain:

$$\frac{\sigma}{E_{ct}} = \left(\frac{\sigma}{E_m} \right) V_m + \left(\frac{\sigma}{E_f} \right) V_f \tag{15.11}$$

or $$E_{ct} = \frac{E_m E_f}{E_f V_m + E_m V_f} \tag{15.12}$$

where E_{ct} is the Young's modulus of the composite under transverse loading (see Example 15.8).

Fig. 15.7 A unidirectional continuous fiber composite under transverse loading

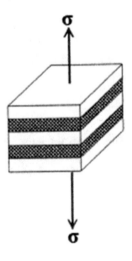

15.2.4 Poisson's Ratio of Composite Material

The Poisson's ratio of a composite material (v_{12}) is given by:

$$v_{12} = -\frac{Stain\,in\,direction\,2}{Strain\,in\,direction\,1} \qquad (15.13)$$

The Poisson's ratio of a fibrous composite can be calculated by:

$$v_{12} = V_m v_m + V_f v_f \qquad (15.14)$$

where the properties are measured in *1–2 plane* (see Example 15.9). It can also be shown that:

$$\frac{v_{12}}{v_{21}} = \frac{E_1}{E_2} \qquad (15.15)$$

where v_{21} is the Poisson's ratio in 2–1 plane; E_1 is the Young's modulus measured in direction 1; and E_2 is the Young's modulus measured in direction 2 (see Example 15.10).

15.2.5 Shear Modulus of Fibrous Composite Materials

The shear modulus for an unidirectional fiber composite in 1–2 plane can be determined by:

$$G_{12} = \frac{G_m G_f}{V_m G_f + V_f G_m} \qquad (15.16)$$

The significance of Eq. 15.16 is illustrated in Examples 15.12 and 15.13.

15.3 Mechanical Behavior of Particulate Composites

A particulate composite (or a particle-reinforced composite), may be either a large-particle composite or a dispersion-strengthened composite (see Chap. 5, Fig. 5.9). Polymers, in general, have low stiffness and low strength but filling with a few weight percent of micro-nano silica, glass, or alumina particles dramatically improves the stiffness, strength, and fracture toughness. Based on rigid-particle assumption, the modulus of elasticity of a particulate composite (E_c) can be related to the modulus of the polymer matrix (E_m) and the volume fraction of the particles (V_p) by the following formula (Fu et al. 2008):

$$E_c = E_m \left(1 + 2.5V_p + 14.1V_p^2\right) \qquad (15.17)$$

The significance of Eq. 15.17 is illustrated in Example 15.14.

The tensile strength (σ_{ut}) of a composite depends on the weakest fracture throughout the material. The tensile strength of a particulate polymer-matrix composite can be related to the inter-particle spacing (D_s) by the following relationship (Young and Beaumont 1977):

$$\sigma_c = \sigma_m + \frac{S}{D_s} \qquad (15.18)$$

where σ_c is the tensile strength of the composite, MPa; σ_m is the tensile strength of the matrix, MPa; and S is a constant that depends on the interface adhesion, MPa-m. The inter-particle spacing (D_s) is given by:

$$D_s = \frac{2d_p\left(1 - V_p\right)}{3V_p} \qquad (15.19)$$

where d_p is the particle diameter, and V_p is the volume fraction of particles (see Example 15.15).

15.4   Calculations – Worked Examples

Example 15.1 Calculating the Critical Fiber Length   It is required to produce a glass/epoxy composite by using a glass fiber with a diameter of 12 μm. The tensile strength of the glass fiber is 3450 MPa, and the adhesive bond strength of epoxy is 20 MPa. Calculate the critical fiber length to produce the composite.

Solution $d = 12\,\mu m = 0.012mm$, $\sigma_f = 3,450MPa$, $\tau_c = 20MPa$, $l_c = ?$

By using Eq. 15.1,

$$\frac{l_c}{d} = \frac{\sigma_f}{2\tau_c} = \frac{3450}{2 \times 20} = 86.25$$

$$l_c = 86.25 \times d = 86.25 \times 0.012 = 1.03mm$$

Example 15.2 Calculating Young's Modulus of GFRP Composite Under Iso-Strain Condition   A glass fiber reinforced polymer (GFRP) composite has 55% by volume of unidirectional continuous glass fibers with a modulus of elasticity of 70 GPa and a tensile strength of 3500 MPa; and an epoxy resin (matrix) with a modulus of 2.5 GPa and a tensile strength of 55 MPa. A tensile stress of 60 MPa is applied in the longitudinal direction over a cross-sectional area of 270 mm² of the composite. Calculate the modulus of elasticity of the composite under longitudinal loading.

Solution $V_f = 55\% = 0.55$, $V_m = 1 - 0.55 = 0.45$,
$E_f = 70\text{GPa}$, $E_m = 2.5\text{GPa}$, $E_{cl} = ?$

By using Eq. 15.6,

$$E_{cl} = E_f V_f + E_m V_m = (70 \times 0.55) + (2.5 \times 0.45) = 39.6\text{GPa}$$

The modulus of elasticity of the GFRP composite under longitudinal loading = 39.6 GPa

Example 15.3 Calculating the Force Carried By the Fiber Phase and by the Matrix Phase   By using the data in Example 15.2, calculate the force carried by the fiber phase and by the matrix phase.

Solution $\sigma = 60\text{MPa}, A_c = 270\text{mm}^2$, $E_f = 70\text{GPa}$, $E_m = 2.5\text{Gpa}$,
$V_f = 0.55$, $V_m = 0.45$, $F_f = ?$, $F_m = ?$

$$F = F_c = \sigma A_c = 60 \times 270 = 16,200\text{N}$$

By using Eq. 15.2

$$F_f + F_m = F_c = 16,200\text{N} \tag{15.3a}$$

By using Eq. 15.7,

$$\frac{F_f}{F_m} = \frac{E_f V_f}{E_m V_m} = \frac{70GPa \times 0.55}{2.5GPa \times 0.45} = 34.22$$

$$\text{or} \quad F_f = 34.22 F_m \tag{15.3b}$$

By simultaneously solving Eqs. 15.3a and 15.3b, we obtain:
    Force carried by the matrix phase = F_m = 460 N, Force carried by fiber phase = F_f = 15,740 N.

Example 15.4 Calculating the Fraction of Force Carried By the Fibers in the Composite   By using the data in Example 15.3, calculate the fraction of force carried by the fibers phase.

Solution $F_m = 460\text{N}, \ F_f = 15{,}740\text{N}$

$$\text{Fraction of force carried by fibers} = \frac{F_f}{F_f + F_m} = \frac{15{,}740}{15{,}740 + 460} = 0.97$$

It means that almost all (97% of) the load acting on the composite is borne by the fibers phase.

Example 15.5 Calculating the Stresses for Each of Fiber and Matrix Phases in a Composite By using the data in Examples 15.2 and 15.3, calculate the stresses for each of the fiber and the matrix phases.

Solution $V_f = 0.55, \ V_m = 0.45, \ F_m = 460\text{N}, \ F_f = 15{,}740\text{N}$

$$\text{Cross-sectional area of fibers} = A_f = A_c V_f = 270 \times 0.55 = 148.5 \ \text{mm}^2 = A_f$$

$$\text{Cross sectional area of matrix} = A_m = A_c V_m = 270 \times 0.45 = 121.5\text{mm}^2$$

$$\text{Stress for fiber phase} = \sigma_f = \frac{F_f}{A_f} = \frac{15740}{148.5} = 106\text{MPa}$$

$$\text{Stress for matrix phase} = \sigma_m = \frac{F_m}{A_m} = \frac{460}{121.5} = 3.78\text{MPa}$$

Example 15.6 Calculating the Strains for Each of the Fiber and Matrix Phases By using the data in Examples 15.2, 15.3, 15.4, and 15.5, calculate the strains for each of the fiber and the matrix phases.

Solution $\sigma_f = 106\text{MPa}, \ \sigma_m = 3.78\text{MPa}, \ E_f = 70\text{GPa} = 70{,}000\text{MPa},$
$E_m = 2.5\text{GPa} = 2500\text{MPa}$

By using the modified form of Eq. 3.7,

$$\text{Strain in the fibers phase} = \varepsilon_f = \frac{\sigma_f}{E_f} = \frac{106}{70{,}000} = 1.51 \times 10^{-3}$$

$$\text{Strain in the matrix phase} = \varepsilon_m = \frac{\sigma_m}{E_m} = \frac{3.78}{2{,}500} = 1.51 \times 10^{-3}$$

These strain results confirm that $\varepsilon_c = \varepsilon_f = \varepsilon_m$.

Example 15.7 Calculating Tensile Strength of a Composite Under Longitudinal Loading By using the data in Example 15.2, calculate the tensile strength of the GFRP composite under longitudinal loading.

Solution $V_f = 0.55, \ V_m = 0.45, \ \sigma_f = 3500\text{MPa}, \ \sigma_m = 55\text{MPa}, \ \sigma_{cl} = ?$

By using Eq. 15.8,

$$\sigma_{cl} = \sigma_f V_f + \sigma_m V_m = (3500 \times 0.55) + (55 \times 0.45) = 1,950 \text{MPa}$$

The tensile strength of the GFRP composite under longitudinal loading = 1950 MPa.

Example 15.8 Computing the Young's Modulus of a Composite Under Transverse Loading By using the data in Example 15.2, calculate the modulus of elasticity of the composite under transverse loading.

Solution $V_f = 0.55$, $V_m = 0.45$, $E_f = 70 \text{GPa}$, $E_m = 2.5 \text{GPa}$, $E_{ct} = ?$

By using Eq. 15.12,

$$E_{ct} = \frac{E_m E_f}{E_f V_m + E_m V_f} = 5.32 \text{GPa}$$

Example 15.9 Calculating the Poisson's Ratio of a Composite Measured in 1–2 Plane A fibrous composite contains 55% by volume of fibers with a Poisson's ratio of 0.25. The Poisson's ratio of matrix is 0.4. The modulus of elasticity of the composite measured in direction 1 is 30 GPa, and that measured in direction 2 is 4 GPa. Calculate the Poisson's ratio of the composite measured in *1–2 plane*.

Solution $V_f = 55\% = 0.55$, $V_m = 1 - 0.55 = 0.45$, $v_f = 0.25$, $v_m = 0.40$, $v_{12} = ?$

By using Eq. 15.14,

$$v_{12} = V_m v_m + V_f v_f = (0.45 \times 0.4) + (0.55 \times 0.25) = 0.32$$

The Poisson's ratio of the composite measured in *1 - 2 plane* = $v_{12} = 0.32$.

Example 15.10 Calculating the Poisson's Ratio of a Composite as Measured in 2–1 Plane By using the data in Example 15.9, calculate the Poisson's ratio of the composite as measured in *2–1 plane*.

Solution $E_1 = 30 \text{GPa}$, $E_2 = 4 \text{GPa}$, $v_{12} = 0.32$, $v_{21} = ?$

By using the modified form of Eq. 15.15,

$$v_{21} = \frac{v_{12} E_2}{E_1} = \frac{0.32 \times 4}{30} = 0.042$$

The Poisson's ratio of the composite as measured in *2–1 plane* = 0.042.

Example 15.11 Calculating the Shear Moduli of Fiber and Matrix in a Composite By using the data in Example 15.2, calculate the in-plane shear modulus for each of the fibers and the matrix phases, if the Poisson's ratios of the fiber and matrix are 0.21 and 0.42, respectively.

Solution $E_f = 70\text{GPa}$, $E_m = 2.5\text{GPa}$, $\nu_f = 0.21$, $\nu_m = 0.42$, $G_f = ?$, $G_m = ?$

By using the re-written form of Eq. 7.5,

$$\text{The shear modulus of the fiber} = G_f = \frac{E_f}{2(1+\nu_f)} = \frac{70}{2(1+0.21)} = 28.92\text{GPa}$$

$$\text{The shear modulus of the matrix} = G_m = \frac{E_m}{2(1+\nu_m)} = \frac{2.5}{2(1+0.42)} = 0.88\text{GPa}$$

Example 15.12 Calculating the In-plane Shear Modulus of a Composite By using the data in Example 15.11, calculate the in-plane shear modulus of the composite.

Solution $V_f = 0.55$, $V_m = 0.45$, $G_f = 28.92\text{GPa}$, $G_m = 0.88\text{GPa}$, $G_{12} = ?$

By using Eq. 15.16,

$$G_{12} = \frac{G_m G_f}{V_m G_f + V_f G_m} = \frac{0.88 \times 28.92}{(0.45 \times 28.92) + (0.55 \times 0.88)} = \frac{25.45}{13 + 0.48} = 1.888\text{GPa}$$

The in-plane shear modulus of the composite = 1.888 GPa.

Example 15.13 Computing Young's Modulus of a Composite By Using Shear Modulus Data A fibrous composite has 63% by volume of unidirectional continuous fibers with a shear modulus of 90 GPa. and a matrix with a shear modulus of 1.4 GPa. The Poisson's ratios of the fibers and the matrix are 0.22 and 0.38, respectively. Calculate the modulus of elasticity of the composite under longitudinal loading.

Solution $V_f = 0.63$, $V_m = 0.37$, $G_f = 90\text{GPa}$, $G_m = 1.4\text{GPa}$,
$\nu_f = 0.22$, $\nu_m = 0.38$, $E_{cl} = ?$

By using Eq. 7.5,

$$E_f = 2G_f(1+\nu_f) = 2 \times 90(1+0.22) = 219.6\text{GPa}$$

$$E_m = 2G_m(1+\nu_m) = 2 \times 1.4(1+0.38) = 3.86\text{GPa}$$

By using Eq. 15.6,

$$E_{cl} = E_f V_f + E_m V_m = (219.6 \times 0.63) + (3.86 \times 0.37) = 139.78 \text{GPa}$$

The modulus of elasticity of the composite under longitudinal loading = 139.78 GPa.

Example 15.14 Calculating the Young's Modulus of Matrix in a Particulate Composite The modulus of elasticity of nylon-6 nano-composite is 2.5 GPa. The composite contains silica particles with a volume fraction of 0.10 and with a particle size of 100 nm. Calculate the Young's modulus of the polymer matrix in the composite.

Solution $E_c = 2.5 \text{GPa}$, $V_p = 0.10$, $E_m = ?$

By using the re-written form of Eq. 15.17,

$$E_m = \frac{E_c}{1 + 2.5V_p + 14.1V_p^2} = \frac{2.5}{1 + (2.5 \times 0.1) + (14.1 \times 0.1^2)} = 1.8 \text{GPa}$$

The Young's modulus of the polymer matrix = 1.8 GPa.

Example 15.15 Calculating the Inter-Particle Spacing and Tensile Strength of a Composite The tensile strength of an epoxy-matrix is 108 MPa. The epoxy-matrix composite contains 50 wt.% glass beads with a particle size of 0.06 mm. Calculate the: (a) inter-particle spacing, and (b) tensile strength of the composite. Take constant S (for interface adhesion) = 1.74 × 10⁻³ MPa•m.

Solution $\sigma_m = 108 \text{MPa}$, $V_p = 50\% = 0.50$, $d_p = 0.06 \text{mm}$,
$S = 1.74 \times 10^{-3} \text{MPa} \times \text{m}$, $\sigma_c = ?$

(a) By using Eq. 15.19,

$$D_s = \frac{2d_p(1 - V_p)}{3V_p} = \frac{2 \times 0.06(1 - 0.5)}{3 \times 0.5} = 0.04 \text{mm}$$

(b) By using Eq. 15.18,

$$\sigma_c = \sigma_m + \frac{S}{D_s} = 108 \text{MPa} + \frac{1.74 \times 10^{-3} \text{MPa} \cdot \text{m}}{0.04 \text{mm}} =$$

$$108 \text{MPa} + \frac{1.74 \text{MPa} \cdot \text{mm}}{0.04 \text{mm}} = 151.5 \text{MPa}$$

The tensile strength of the composite = 151.5 MPa.

Questions and Problems

15.1. Draw the classification chart of composite materials.

15.2. What are the roles of matrix material and reinforcing material in a composite?

15.3. Draw diagrams showing various composites based on reinforcing material structures.

15.4. Draw diagrams showing various fibrous composites.

15.5. Why are CFRP composites extensively used in aerospace structures?

15.6. Differentiate between iso-strain condition and iso-stress condition with the aid of diagrams.

15.7. Derive Eq. 15.6.

15.8. The Young's modulus of aluminum alloys is 70 GPa and the density is 2.73 g/cm^3. Compute and compare the specific moduli of aluminum alloys and CFRP.

15.9. The tensile strength of a glass fiber is 3400 MPa, and its diameter is 13 μm. The adhesive bond strength of epoxy is 22 MPa. Calculate the critical fiber length to produce the composite.

15.10. A GFRP composite has 45% by volume of unidirectional continuous glass fibers with a Young's modulus of 70 GPa and a tensile strength of 3480 MPa; and an epoxy matrix with a modulus of 2.4 GPa and a tensile strength of 54 MPa. A tensile stress of 70 MPa is applied in the longitudinal direction over an area of 280 mm^2 of the composite. Calculate the (a) Young's modulus of the composite under longitudinal loading, (b) force carried by the fibers and by the matrix, (c) stresses for each of the fibers phase and the matrix, (d) strains for each of the fibers phase and the matrix, (e) tensile strength of the composite under longitudinal loading, and (f) Young's modulus the composite under transverse loading.

15.11. A fibrous composite contains 60% by volume of fibers with a Poisson's ratio of 0.26. The Poisson's ratio of the matrix is 0.41. The modulus of elasticity of the composite measured in direction 1 is 32 GPa, and that measured in direction 2 is 4.2 GPa. Calculate the Poisson's ratio of the composite measured in *1–2 plane.*

15.12. By using the data in *Problem 15.10*, calculate the in-plane shear modulus for each of the fibers and the matrix phases. The Poisson's ratios of the fiber is 0.22 and that of matrix is 0.44.

15.13. A fibrous composite has 57% by volume of unidirectional continuous fibers with a shear modulus of 92 GPa. and a matrix with a shear modulus of 1.5 GPa. The Poisson's ratios of the fibers and the matrix are 0.20 and 0.36, respectively. Calculate the modulus of elasticity of the composite under longitudinal loading.

15.14. An epoxy-matrix composite contains 40 wt.% glass beads with a particle size of 0.05 mm. The tensile strength of the matrix is 107 MPa. Calculate the: (a) inter-particle spacing, and (b) tensile strength of the composite. Take the constant $S = 1.5 \times 10^{-3}$ MPa•m.

15.15. The Young's modulus of nylon-6 nano-composite is 2.5 GPa. The composite contains silica particles with a volume fraction of 0.30 and with a particle size of 90 nm. Calculate the Young's modulus of the polymer matrix in the composite.

15.16. (MCQs). Encircle the correct answers for each of the following questions. (a) Which of the following materials possesses the highest specific modulus? (i) GFRP, (ii) CFRP, (iii) steel, (iv) aluminum (b) What should be the critical fiber length-to-diameter ratio for a fibrous composite? (i) 100–150, (ii) 20–110, (iii) 10–80, (iv) 30–140 (c) Which type of composite corresponds to concrete? (i) particulate, (ii) fibrous, (iii) laminate, (iv) none

References

Farley GL, Jones RM (1992) Crushing characteristics of continuous fiber-reinforced composite tubes. J Compos Mater 26:37–50

Fu S-Y, Feng X-Q, Lauke B, Mai Y-W (2008) Effects of particle size, particles/matrix interface adhesion and particle loading on mechanical properties of particulate-polymer composites. Compos Part B 39:933–961

Gay D (2015) Composite materials: design and applications. CRC Press, Boca Raton, USA

Kaw AK (2006) Mechanics of composite materials, 2nd edn. CRC Press, Boca Raton, FL

Young RJ, Beaumont PWR (1977) Effect of composition on fracture of silica particle – filled epoxy resin composites. J Mater Sci 12:684–692

Answers to Problems

Chapter 1

P1.7. $FoS = 0.45$, Since FoS < 1, the component is unsafe for the application.
1.8(MCQs). (a) *ii*, (b) *i*, (c) *iv*, (d) *ii*, (e) *iii*, (f) *i*, (g) *iv*, (h) iii, (i) iii, (j) iv

Chapter 2

P2.9.   0.4086 nm
P2.11. 0.673 MPa
P2.13. 2.487 MPa
P2.15. (a) 1.3652, (b) 35 mm, (c) 74.5

Chapter 3

3.1(MCQs). (a) iii, (b) i, (c) iv, (d) ii, (e) i, (f) iv, (g) iv, (h) i
P3.7.   (a) 636.5 MPa, (b) 0.093
P3.9. $\sigma_{ut} = 255$ MPa, $\sigma_{break} = 255$ MPa
P3.11. 230.72

© The Author(s), under exclusive license to Springer Nature Switzerland AG 2022   283
Z. Huda, *Mechanical Behavior of Materials*, Mechanical Engineering Series,
https://doi.org/10.1007/978-3-030-84927-6

Chapter 4

4.1(MCQs). (a) ii, (b) i, (c) iii, (d) i, (e) ii
P4.9.   142 nm
P4.11. $3.3 \times 10^8 \ cm^{-2}$
P4.13. 30.92 MPa

Chapter 5

P5.15. The 2024-T81 *Al* alloy has 3 times better specific strength than that of carbon steel.
P5.17(MCQs).
(a) i, (b) iii, (c) ii, (d) ii, (e) i, (f) i, (g) iii, (h) iii, (i) iii, (j) iv, (k) i, (*l*) iii, (m) i,
(n) i, (o) ii, (p) ii, (q) i

Chapter 6

6.1(MCQs). (a) iii, (b) i, (c) iii, (d) ii, (e) i
P6.3. (a) 0.162, (b) 1058.8 MPa
P6.5. % CW = 24.2
P6.7. $H = 604.2$ MPa

Chapter 7

P7.5.   0.268
P7.7.   For titanium: $U_r = 0.844$ MJ/m^3, For steel: $U_r = 1.55$ MJ/m^3; Hence steel is
        selected.
P7.9. $\varepsilon_x = 3.9 \times 10^{-4}$ $\varepsilon_y = 1.61 \times 10^{-3}$ $\varepsilon_z = -8.29 \times 10^{-4}$
P7.11. $\gamma_{xy} = 1.52 \times 10^{-3}$ $\gamma_{yz} = 1.41 \times 10^{-3}$ $\gamma_{zx} = 1.74 \times 10^{-3}$
P7.13. $\varepsilon_{x(T)} = 14.7 \times 10^{-4}$ $\varepsilon_{y(T)} = 9.82 \times 10^{-4}$ $\varepsilon_{z(T)} = 4.94 \times 10^{-4}$
7.14(MCQs).   (a) ii   (b) iv   (c) i   (d) iii   (e) iii

Chapter 8

P8.5.   Direct stress = 47.37 MPa, Shear stress = 98.5 MPa
P8.7. $\tau_{xy} = \tau_{yx} = 56.79$ MPa
P8.9.   (a) $\varepsilon_1 = 7.02 \times 10^{-3}$, $\varepsilon_2 = -4.12 \times 10^{-3}$, $\varepsilon_3 = -3.03 \times 10^{-3}$ (b) $\theta_n = 14.5°$ (CCW)
P8.11. 2.9×10^{-3}

Chapter 9

P9.7. (a) 1.76 mm, (b) –0.089 mm, (c) –0.0343

P9.9. $m = \text{slope} = \dfrac{\Delta \log \sigma}{\Delta \log \dot{\varepsilon}} = \dfrac{\log \sigma_2 - \log \sigma_1}{\log \dot{\varepsilon}_2 - \log \dot{\varepsilon}_1}$

P9.11. (a) 80 MPa, (b) $\sigma_1' = 190$ MPa, $\sigma_2' = -280$ MPa, $\sigma_3' = 90$ MPa

P9.13. $m = 0.7$; yes

9.15(MCQs). (a) ii, (b) iv, (c) iii, (d) i, (e) i, (f) iii.

Chapter 10

P10.3. 93.9 MPa

P10.5. 47.62 N·m

P10.7. 32.5 mm

P10.9. (a) 14,700 N·m/rad., (b) 68×10^{-6} rad/N·m

Chapter 11

P11.5. $|\sigma_1| = 248.5$ MPa, $|\sigma_2| = 181.5$ MPa, $|\sigma_1 - \sigma_2| = 97$ MPa, Failure will not occur.

P11.7. 222.8 MPa < 450 MPa, Failure will not occur.

P11.9. $D = 38$ mm or 40 mm.

11.11(MCQs).   (a) iii, (b) ii, (c) i, (d) ii, (e) i

Chapter 12

12.1(MCQs). (a) i, (b) iii, (c) ii, (d) iii, (e) i, (f) iv, (g) ii, (h) ii, (i) iii

P12.5. $K_s = 25.8$

P12.7. $\sigma_c = 608$ MPa, Design stress $\cong 580$ MPa

P12.9. $K_I = 8.74$ MPa $\sqrt{m}$, the design is safe.

Chapter 13

P13.7. (a) 0, (b) 170 MPa, (c) 85 MPa, (d) – 1, (e) $\propto$

P13.9. 180 MPa

P13.11. 0.12

P13.13(MCQs). (a) ii, (b) iii, (c) i. (d) iii, (e) i, (f) ii.

Chapter 14

P14.7.   1,830 h
P14.9.   30.7 years
P14.11. 115 MPa.
14.12(MCQs). (a) ii, (b) i, (c) iv, (d) iii.

Chapter 15

P15.9.   1.00 mm
P15.11. 0.324
P15.13. 127.6 GPa
P15.15. 0.828 GPa
P15.16(MCQs). (a) ii, (b) iv, (c) i

Index

A

Accidental loads, 8
Activation energies of creep
   defined, 257
   determination, 257, 261
Aging, *see* Precipitation hardening
Aircraft structural materials, 89–93, 95, 239
Air hardening steels, 89
AISI-1020, 46, 56, 57
AISI 304N, 76
Allowable stress (working stress) in design, 12
Alloyed cast iron, 83, 85
Alloying, *see* Solid-solution strengthening
Alloys, 48–51, 54, 63–78, 81–83, 86–95, 102,
      103, 112, 117, 121, 187, 223, 228,
      237, 243, 258, 263, 264
Alloy steels
   Hadfield manganese steel, 89, 90
   maraging steels, 70, 89, 90
   stainless steels, 89, 90, 112
   4140 steel, 89
   4340 steel, 89, 112, 119
Alpha brass, 94
Alpha-iron, 17
Alternating stress cycle, 233–236, 242, 243,
      249, 250
Alumina (Al_2O_3), 46, 96–98, 208, 273
Aluminum and alloys, 50–51, 67, 69, 82, 83,
      91–93, 102, 103, 127, 223, 228,
      236, 243
American Iron and Steel Institute (AISI), 46,
      56, 57, 68, 69, 86, 87
American Society for Testing and Materials
      (ASTM), 42
Amorphous solid, 15

Amplitude, of stress, 235–238, 242, 243
Amplitude ratio, 235, 236, 243
Angle of rotation, 145, 148,
      152–154, 161–163
Angle of twist, 41, 54, 190, 192, 193, 195–198
Anisotropic materials, 46, 119, 140
Annealing of steel, 73
Austempered ductile iron (ADI), 83, 85, 104
Austenite (γ-iron), 87, 90, 91
Austenitic stainless steels (ASS), 68, 69, 90
Axial strain, 60, 120, 122, 130
Axial stress, 233, 234

B

Barium titanate, 97, 106
Bauschinger effect, 166, 168, 186, 188,
      201, 213
Beach marks in fatigue failure, 233
Beams, 6, 91, 144, 170–171, 180–182, 186,
      235, 250, 253
Bending of beams
   analysis, 169–171
   curvature, 169–170
   radius of curvature, 169–170
   symmetrical bending, 170–171
Bending stress, 234, 242
Biaxial stresses, 201
Biomaterial, 81, 95
Body-centered cubic (BCC) structure, 17–20,
      26, 30, 69, 78, 90
Bonding in solids, 15–16
Bones, 11
Boron carbide (B_4C), 96
Boron nitride (BN), 96

Brale indenter, 48
Branching in polymers, 97
Brass, 8, 34, 49, 51, 83, 94, 127
Bridges
    failure of, 6, 215
    fatigue life of, 7, 239, 248, 250
Brinell hardness test, 47, 48, 58
Brittle behavior, 7–10
    effects of cracks on, 215, 216, 221
    multiaxial criteria for, 8, 12
    in notch fatigue, 238, 239
Brittle fracture, 7–10, 201, 215, 217
    techniques against, 238–239
Bronze, 83, 94, 104, 127, 264
Bulk modulus, 40, 46, 118, 119, 124–128,
        138, 144
Burger's vector, 23, 24, 36

C
Cabin window corner radii, 220, 230
Carbon fiber reinforced polymer (CFRP)
        composite, 99, 100, 267, 281
Carbon steels, 49, 83, 86–88, 100, 102, 104,
        112, 127, 210, 213, 223, 236
    AISI-SAE designations for, 89
    mechanical properties of, 88
    microstructures of, 87, 88
Cartridge brass, 94
Casting, 12, 39, 64, 85, 92, 258
Cast irons
    characteristics and applications of, 83
    mechanical properties of, 85, 86
    types of, 83–85
Cemented carbides, 50, 100
Cementite, 84, 85, 87
Ceramic matrix composites (CMC), 100, 267
Ceramics, 16, 81, 96–98, 100, 104, 227, 228,
        230, 258, 267
    clay products, 96
    concrete, 267
    engineering (advanced), 97
Chain of molecules, 97
Challenger, STS-51L space shuttle, 215
Charpy V-notch test, 52, 59, 61
Chromium in steel, 83, 88
Circular (embedded) cracks, 222, 229, 230
Circular shafts, torsion of, 189, 195, 197
Clay, 96
Cleavage, 7, 10
Climb (dislocation), 253, 254, 264
Coefficient of thermal expansion (α), 95, 128,
        139, 144, 269
Cold work (CW), 27–28, 34, 54, 67–69, 75,
        76, 90, 92, 94, 114, 116, 117

Compacted graphite (CG) cast iron, 83, 85
Complex strain, 152–153, 162
Complex stress, 3, 145–148, 152, 155, 162
Composite materials, 99, 100, 267–281
    applications, 267, 269
    classification, 267, 268
    definition, 267
    fibrous composites, 100, 267–273
    mechanical behavior of, 100, 267–281
    particulate composites, 100, 267,
        270, 273–274
Crack initiation, 8, 11, 85, 233
Crack propagation, 11, 222, 230, 233
Crack termination, 11, 233
Creep, 3, 6, 7, 40, 53, 54, 95, 96, 103, 201,
        215, 253–264
    curve (creep curve), 253–256
    deformation, 3, 6, 7, 13, 253, 254
    failure, 253–255
    primary creep, 253–255
    rate of creep, 256–257, 260–264
    secondary (steady-state) creep, 253, 256
    tertiary creep, 253, 255
    testing (creep testing), 253–255
Creep-limited alloy design, 258
Critical crack size (a_c), 223, 229, 238,
        249, 250
Critical fiber length (l_c), 267, 270, 274
Critical resolved shear stress τ_{crss}, 29, 34–36
Critical stress intensity factor K_c, 221,
        222, 230
Crystal imperfections (defects), 22–24
Crystallographic directions, 20, 21, 32, 36
Crystallographic planes, 7, 15, 20–22, 24, 25,
        32, 33, 36
Crystallography, 15–21
Crystal structure, 3, 15–20, 23–25, 30, 32, 72
    BCC, 17–19, 26, 30
    FCC, 18–20, 25, 30
    HCP, 19, 20, 26, 27, 30, 32
Crystal structure properties, 17, 19
Crystal systems, 15–17, 21
Cumulative damage, 239–240, 246, 250
Curvature, 169–171, 180, 181, 186, 188,
        218–219, 226, 227, 230
Cyclic stress-strain behavior, 233–235

D
Deflection in beam, 169, 180, 187
Deformation
    characteristics of the various types
        of, 23–24
    creep, 3, 6–7, 14, 253, 254
    elastic type, 4–6, 14, 113–114, 125

plastic type, 3, 4, 6, 8–11, 13, 22–29, 36, 45, 47, 63, 78, 111, 113–114, 167, 169–171, 179, 190, 201, 213
  by slip, 3, 15, 22, 24–26, 28, 34, 36
  by twin, 25–27
Deformation behavior models, 113–114
Degree of crystallinity, 98
Degree of polymerization (DP), 98, 103
Design
  creep, 260
  defined, 12
  fatigue, 238–239
  material selection, 11–12, 224
  safety factors, 12–13, 204, 205, 210, 212, 228
Design philosophy of fracture mechanics, 224, 228
Deviatoric stress, 174–175, 184, 186–188
Diamond crystal structure, 16
Die swell, 129, 130, 140
Dislocation climb (in creep), 253, 254, 264
Dislocation density (ρ_D), 23, 25, 27, 35, 36, 67, 68, 74–76, 78
Dislocation motion, plastic deformation by, 15, 28
Dislocation piling, 65, 66
Dislocations, 3, 15, 22–25, 28, 29, 35, 36, 63–70, 72, 74–78, 168, 253, 254, 258, 264
Dispersion strengthening, 64, 71, 72, 77, 78
Draft, 27, 34, 36
Ductile behavior in a tension test, 42–43
Ductile fracture, 7–10, 13, 201
Ductile iron, 83, 85
Ductility
  engineering measures of, 46, 52, 53
  and necking, 45, 166–168
  percent elongation, 45, 57
  percent reduction in area, 45
Duplex stainless steel (DSS), 90, 91

E

Edge dislocation, 23–25, 63, 64, 253
Effective strain, 177, 184, 187
Effective stress, 177, 185, 187
Elastic constants, 46, 119, 121, 124–128, 138, 140
Elastic deformation
  bulk modulus, 46, 119, 124–128, 138
  hydrostatic stress, 124, 125
  thermal strains, 128, 129
  volume change in elastic deformation, 125, 126, 138, 140
  volumetric strain, 124–126, 138

Elastic limit, 4, 45, 113, 119
Elastic, linear-hardening stress-strain relationship, 111–113
Elastic modulus (Young's modulus)
  under longitudinal loading, 271, 275, 278, 279
  under transverse loading, 274, 277, 280
Elastic, perfectly plastic stress-strain relationship, 112, 118
Elastic strain, 4, 44, 45, 57, 111–113, 118, 121, 128–129, 165, 254
Elastic strains, Hooke's law for, 6, 124
Elastic, work-hardening stress-strain relationship, 112, 201, 213
Elastomers, 129
Electronic materials, 81, 101
Elliptical cracks, 217
Elongation, percent, 45, 57, 60
Embrittlement, 7, 201
Endurance limit, 236, 237
Energy, impact, 9, 10, 13, 52–54, 59, 61
Engineering ceramics, 97
Engineering components
  materials selection for, 11, 224
Engineering design, 186
Engineering materials
  classes and examples of, 81, 82
  general characteristics of, 40, 46
Engineering stress and strain, 42, 43, 55, 56, 109, 110, 115, 165, 167
Engineering stress-strain properties
  breaking strength, 45, 166
  ductility, 45
  elastic limit, 4, 45, 119
  elastic (Young's) modulus, 4, 45, 56, 111–112, 115
  elongation, 43, 45, 46, 57, 58, 166
  lower yield point, 165
  necking behavior and ductility, 45, 166
  proportional limit, 44, 45, 56, 57
  reduction in area, 45, 46, 168
  strain hardening, 64, 67, 68, 112, 116, 166, 168
  tangent modulus, 111–112, 115
  from tension, 45, 46, 144, 165
  ultimate tensile strength, 44, 45, 48, 166, 170, 238
  upper yield point, 165–166
  yielding, 45, 121, 166, 168
  yield strength, 45, 46, 57, 166, 168
Environmental assisted cracking (EAC), 7, 8
Epoxy, 99, 104, 267, 274, 280
Etched sample, 67, 74, 78
Extensometer, 42
Extrinsic semiconductors, 101

F
Face-centered cubic (FCC) structure, 18, 19
Factor of safety *(FoS)*, 12–13,
        204–207, 210–213
Failure criteria
    brittle fracture criteria, 7–10, 201, 215, 217
    yield criteria, 113–114
Failure envelope, for Mohr's circle, 149–150
Failure surface, 208–210
Fatigue
    crack initiations in, 11, 85, 233
    cyclic loading, 235, 236, 241
    definitions for, 11
    design against, 10, 216
    fatigue damage, 239
    fatigue limit behavior, 236, 238, 243, 244
    fatigue testing, 236–237
    fracture mechanics approach, 3, 215–229
    HCF and LCF, 237
    life estimates with, 236–237,
        239–243, 245–250
    mean stresses, 235, 236, 238, 242, 244
    Miner's rule, 239–240
    notch effects, 244, 250
    residual stress effects, 238
    S-N curves, 236, 243, 245
Fatigue crack growth
    fatigue crack-growth rate, 240–242, 247,
        249, 250
    Paris equation, 249–250
Fatigue failure
    designing to avoid, 201, 238, 239
    stages in, 11, 12, 234
    surface residual stresses, 238
Fatigue failure prevention/avoidance, 201
Fatigue limits, 236, 238, 244, 245
Fatigue strength, 39, 85, 86, 91, 105, 236–239,
        243, 249
Fatigue stress concentration factor, 239, 244
Fatigue testing, 236–237
Ferrite (α-iron), 91
Ferritic stainless steel (FSS), 90
Ferro-electric material, 97, 106
Ferrous alloys, 83, 86
Fiberglass, 100, 267, 276, 280
Fibrous composites
    E calculation under longitudinal loading,
        271, 275, 278, 279
    E calculation under transverse loading,
        272, 277
    random discontinuous fibers, 267, 268
    unidirectional continuous fibers, 267,
        270–273, 278, 280

unidirectional discontinuous fibers,
        267, 268
Flaw shape factor (geometric factor), 216, 224,
        228, 229, 239
Fluctuating stress cycle, 233, 235
Fracture
    brittle, 7–10, 201, 215, 217
    cleavage, 7, 10
    of cracked members, 221, 224, 225, 240
    intergranular, 9
    modes, 10, 220–221
    for static and impact loading, 8, 10
    transgranular, 7, 9
    types of, 8, 9, 201
Fracture mechanics
    application to design a test method, 224
    application to design stress, 224
    application to predict design safety, 224
    critical stress intensity factor (K_c), 221, 222
    for fatigue crack growth, 240–242, 247
    plane strain fracture toughness (K_{IC}), 221,
        222, 228, 231
    stress concentration factor, 218–220, 226,
        230, 239, 244, 250
    stress intensity factor K, 241
Fracture strength, 57
Fracture surface, 8–10
Fracture toughness, 85, 105, 221, 223, 224,
        228–231, 273
Fully plastic yielding, 165

G
Gage length, 42, 43, 109, 169, 258
Gamma iron, 18
Gas-turbine (GT) engines, 95, 253, 258
Generalized Hooke's law, 124
Generalized plane stress, 150–151, 158, 163
Geometric factor in fracture mechanics, 221
GFRP composite, 267, 275–277, 280
GLARE composite, 267, 269
Glass, 9, 53, 81, 96–98, 100, 127, 223, 224,
        226, 230, 267, 269, 273, 274,
        279, 280
Goodman diagram, 238, 250
Goodman equation, 238
Grain boundaries, 7, 64–67, 77, 78
Grain-boundary strengthening, 64–67, 77, 78
Grained microstructure, 64–65
Grain oriented electrical steel (GOES),
        21, 91, 104
Graphite, 16, 83–85, 101, 102, 267
Gray cast iron, 84

Griffith, A.A., 216–218, 226, 230, 250
Griffith's crack theory, 216–218, 230

H

Hall-Petch relationship, 65–67, 72, 73, 78
Hardening, 47, 64, 66–68, 77, 78, 89,
        112–113, 116, 118, 166, 168,
        201, 213
Hardness, 39, 40, 47–53, 58–60, 67, 68, 83,
        85–90, 102, 105
Hardness correlations and conversions, 52
Hardness tests
     Brinell hardness test, 47–49, 58, 60, 90
     Rockwell hardness test, 47–50, 59
     Vickers hardness test, 47, 50–51, 58–60
Havilland Comet aircraft crash, 218
Heating, ventilation, and air-conditioning
        (HVAC) system, 94
Heat treatment, 47, 70, 85–90, 92–93, 102,
        104, 258
Hexagonal close-packed (HCP) crystal
        structure, 17, 19–20
High-carbon steels, 87, 88
High cycle fatigue (HCF), 237, 249
High density polyethylene (HDPE), 46,
        98, 99, 106
High-temperature creep, 6, 54, 93, 95, 253
Hooke's law, 6, 124
Hot working (HW), 114, 118
Hybrid composites, 267, 269
Hydrogen embrittlement (HE), 7, 201
Hydrostatic stress, 124–127, 138,
        174–175, 184–188

I

Impact energy tests, 52–53, 59, 61
Impact loading, fracture under, 8
Impurity (interstitial, substitutional),
        69, 70, 78
Indentation hardness, 47
Inter-granular (IG) fracture, 7, 14
Intermetallic compounds, 70
Internal combustion (IC) engine, 84
Interstitial solid solution, 69, 70, 78
Intrinsic semiconductors, 101
Irons, cast, *see* Cast irons
Iso-strain condition, 270, 271, 275, 280
Iso-stress condition, 272, 280
Isotropic materials, 47, 113, 119, 121, 129,
        141, 152, 160, 163
Izod impact test, 52
Izod tests, 52

K

Kevlar, 100
Knoop hardness number (KHN), 51, 52, 59
Knoop hardness testing, 47, 51, 59

L

Laminated composites, 100, 267, 269, 281
Larson–Miller parameter (LMP), 257–258,
        263, 264
Lateral strain, 60, 120, 122, 123, 130, 131, 144
Lattice plane and site, 21
Levy-Mises flow rule, 175–176
Liberty Ships failures, 215
Lignin, 267
Linear hardening, 113, 118
Line defects (dislocations), 15, 23
Liquid metal embrittlement, 7
Loading modes: I, II, & III, 221, 229, 230
Low-alloy high-strength (LAHS) steels,
        86, 89, 105
Low-carbon steel, 87–90, 102
Low-density polyethylene (LDPE),
        98, 99, 106
Lower yield point (LYP), 165, 166
Low-temperature creep, 6

M

Malleable iron, 50, 83, 85
Manganese in steel, 86, 88–90, 92
Maraging steels, 70, 89, 90
Martensite, 90, 91, 105
Martensitic stainless steel (MSS), 90, 105
Materials selection, 11–13, 224
Maximum normal stress fracture criterion, 202
Maximum principal normal stress theory/
        Rankine theory, 201, 202
Maximum shear stress, 150–152, 162, 192,
        193, 197, 198, 201–206, 213
Maximum shear stress (MSS) theory (Tresca
        theory), 201–205, 213
Mean stress, 235, 236, 238, 242, 244, 249, 250
Mechanical behavior of materials, 3, 13, 15,
        21, 22, 40, 47
Mechanical testing
     creep testing, 253–255, 259, 260, 264
     fatigue tests, 236–237
     hardness tests, 47–52, 58–60
     notch-impact tests, 52
     tension/tensile tests, 42–47, 57–60, 110,
        111, 114, 115, 118, 130, 131,
        166–167, 172, 178
Medium-carbon steels, 87, 102

Metal matrix composites (MMCs), 100, 267
Metals
    ferrous metals, 83
    nonferrous metals, 83, 91–95
    strengthening methods for, 63
Microstructures, 3, 39, 64–65, 70, 72, 76–78,
        84, 85, 87–88, 90–92, 95, 105, 106,
        177, 178, 258
Micro-void coalescence, *see* Dimpled rupture
Mild steel, 8, 26, 42, 46, 50, 53, 87, 105, 114,
        118, 119, 121, 132, 135, 139, 141,
        165, 166, 177, 186, 208–210
Miller indices, 20–22, 32, 33, 36
Miner's law diagram, 240
Miner's law of cumulative damage, 239–240
Mixed dislocation, 23–24
Models, *see* Deformation models
Modulus of elasticity, *see* Young's modulus
Modulus of rigidity, *see* Shear modulus
Mohr's circle, 143, 149–150, 156–158, 162
Molybdenum in steel, 88, 90
Multiaxial stress effects, 10, 201
    See also Three-dimensional stress-strain
Muntz metal, 94

N
Naval brass, 112
Necking, 166
Nickel-base superalloys, 70, 95, 258
Nitrided steels hardness, 50
Nitrides, 51, 96
Nodular cast iron, 84
Nominal stresses, 218
Nonferrous metals
    aluminum alloys, 91–93
    copper alloys, 93–94
    superalloys, 95
    titanium alloys, 95
Non-grain oriented electrical steels
        (NGOES), 91
Normal stress, 29, 122–124, 126, 133, 136,
        138, 140–141, 145, 148–152, 154,
        155, 157–160, 162, 163, 174, 177,
        183, 187, 188, 201–203, 208, 209,
        212, 229
Notched specimens, 52, 239, 244
Notch effects in fatigue, 244, 250
Notch-free specimen, 239
Notch-impact tests, 52
Notch sensitivity factor, 239, 244, 250
Numerical integration (for crack growth), 3,
        216, 240–242, 247, 250, 255
Nylons, 281

O
Opening loading mode (Mode I), 220–221,
        227, 229–231
Orthotropic materials, 46
Oxide-dispersion-strengthened (ODS)
        alloys, 72
Oxides, 96, 223, 228, 257, 258, 261

P
Paris equation, 247, 249
Particulate composites, 100, 267, 268,
        273–274, 279
Pearlite, 85, 87, 88
Percent elongation, 45, 57, 60
Percent reduction in area (%RA), 45, 46
Perfect crystals, 22, 36
Perfectly plastic stress-strain curve,
        111, 165–166
    relationship, 68, 112
    See also Elastic-perfectly plastic
        stress-strain
Plain-carbon steels, 86
Plain strain, 222, 228, 231, 247
Plane strain fracture toughness, 221, 222,
        228, 231
Plane stress, 144–147, 149–151, 158, 162,
        163, 172–175, 183, 186, 203, 204,
        206–208, 210, 212
Plastic deformation, 3–10, 13, 15, 24, 25,
        27–29, 45, 47, 63, 78, 111,
        113–114, 118, 165, 168–171, 177,
        201, 213
Plasticity, *see* Plastic deformation
Plastic modulus, 111, 115, 116, 118
Plastics, *see* Polymers
Plastic strain, 111, 112, 165, 166, 169
    See also Plastic deformation
Point defects, 23
Poisson's ratio, 46, 47, 60, 119–121, 128, 131,
        138, 141, 270, 273, 277, 278, 280
Polar moment of inertia (J), 189
Polyethylene (PE), 42, 46, 97–99, 104, 121
Polyethylene terephthalate (PET), 98, 99
Polymerization, 97, 98, 103
Polymer matrix composites (PMC), 100, 267,
        274, 279, 281
Polymers
    thermoplastics, 98, 100, 104
    thermosetting plastics, 99, 104
Polypropylene (PP), 98, 99, 103, 106, 223
Polystyrene (PS), 223
Polyvinyl chloride (PVC), 54, 98, 99, 106, 223
Potential energy of plate, 217, 225, 231

Power (motor power/shaft power), 191
Power-hardening stress–strain
    relationship, 112
Precipitate, coherent & non-coherent,
    70–72, 77, 78
Precipitation hardening, 258
Pressure vessels, 134–136, 143, 144
Primary stage of creep, 253–256, 264
Principal axes, 145, 150, 158, 159, 163
Principal normal stresses, 145, 148–152, 154,
    155, 157–159, 162, 163, 174, 177,
    183, 187, 188, 201–203, 208, 212
Principal shear stresses, 149, 150, 152, 156,
    157, 160, 162
Principal strains, 152–153, 172–173, 177, 182,
    183, 187
Principal stresses
    maximum shear stress, 151, 152, 159, 201
Proportional limit, 42, 44, 45, 56, 57, 60

Q

Quartz crystal structure, 15, 16
Quartz fiber reinforced plastic (QFRP)
    composite, 269
Quenching and tempering, 90, 93

R

Radius of curvature, 169–170, 180, 181, 187,
    188, 218, 219, 226, 227, 230
Range of stress, 239
Real crystal, 22
Reduction in area, 8, 45, 46, 88, 102, 166
Refractories, 96
Remaining fatigue lifetime of bridge
    (calculation), 246, 250
Repeating stress cycle, 11, 233, 235
Repeating unit in polymers, 97, 98
Residual stresses and strains, 238
Resilience, 46, 59, 119, 121, 131, 132,
    141, 142
Rigid/linear strain hardening deformation
    model, 113
Rigid/perfectly plastic deformation model, 113
Rockwell hardness test, 47–50, 59
Rolling, 6, 21, 27–28, 34, 67, 68, 85, 89, 91,
    120, 165
Rotating-bending fatigue testing, 237
Rubbers, 127, 129, 138
Rupture in creep, 253, 263, 264

S

SAE steel nomenclature, 86
Safety, 3, 12–13, 91, 205–207, 210, 212, 213,
    224, 228
Safety factor in design, 12–13, 205, 206,
    210, 212
Screw dislocation, 23–25, 36
Secondary stage of creep, 253, 255, 264
Semiconductor fabrication, 22
Semiconductors, 22, 81, 101
Shearing loading mode (*Mode II*), 220,
    229, 230
Shear modulus, 40–42, 46, 55, 119, 121,
    124, 137, 139, 269, 270, 273,
    278, 280
Sheet metal forming, application of plasticity,
    3, 6, 165, 172–174
Ship structures, cracks in, 215
Silica (SiO_2), 15, 16, 96–98, 223, 226, 273,
    279, 281
Silica glasses, *see* Glass
Silicon (polycrystalline silicon), 21, 22, 83,
    88, 90, 92, 94, 101
Silicon carbide (SiC), 96
Single crystals, 15, 28–29, 34, 36, 66,
    78, 95, 258
Sintered alumina powder (SAP), 72
Slip (in crystals), 15, 24–26, 29, 63
S-N (stress *vs.* fatigue life) curves, 236,
    243, 245
Solidification, 23, 39, 64, 65, 258
Solid-solution strengthening, 64,
    69–70, 77
Solution heat treatment, *see* Precipitation
    hardening
Space shuttles, 97, 215
Specific modulus, 267, 281
Specific strength, 93, 95, 100, 102, 103,
    105, 286
Specific surface energy, 217, 218, 224,
    226, 230
Specimens, test
    for fatigue, 236–237, 239, 244
    for notch-impact, 52, 239, 244
    for tension, 42, 165
Spheroidal graphitic (SG) iron, 83–85,
    101, 105
Stainless steels, 54, 68, 69, 76, 89, 90, 112,
    116, 127, 259, 264
Static loading, 218, 226, 227, 230
Steady-state creep, 256, 259, 260, 264

Steels
  as-quenched, 90
  carbon, 49, 83, 86–88, 100, 102, 104, 105,
      112, 127, 210, 213, 223, 236, 284
  LAHS, 86, 89, 105
  mild, 8, 26, 42, 46, 50, 53, 87, 105, 114,
      118, 121, 132, 135, 139, 141, 165,
      166, 177, 186, 208–210
  numbering system, 86
  plain-carbon, 86
  quenching and tempering, 90
  stainless, 54, 68, 69, 76, 89, 90, 112, 116,
      127, 259, 264
  tool, 90
Stiffness, 4, 39, 45, 46, 95, 119, 191, 192, 196,
      198, 270–273
Strain
  complex states of, 3, 143–163
  engineering strain, 42, 43, 55, 58, 60, 115,
      118, 165
  principal strains, 143–163, 172–173, 177,
      182, 183, 187
  strain gage, 43, 169
  transformation of axes, 145
  true strain, 27, 34, 36, 60, 109–112,
      114–118, 167, 169, 179
Strain hardening, 64, 67–69, 77, 78, 112, 116,
      166, 168
Strain hardening exponent, 68
Strain rate sensitivity index, 177, 178,
      186, 187
Strain ratio, 60, 173, 176, 183, 184, 186, 187
Strength coefficient, 40, 68, 112, 177
Strengthening mechanisms for metals, 63–78
Strength, theoretical, 22, 36
Stress
  basic formulas, 13
  components of, 122, 145, 149
  direct stress, 137, 138, 145–148, 153, 155,
      158, 162
  engineering stress, 42, 43, 55, 60, 109,
      115, 118, 165, 166
  generalized plane stress, 150, 151,
      157, 163
  Mohr's circle for, 143, 149, 150, 157
  nominal type, 218
  plane stress, 144–147, 150, 151, 158, 162,
      163, 172–175, 183, 187, 203, 204,
      206–208, 210, 212
  in pressure vessels, 134, 135, 143, 144
  principal stresses, 143–163
  residual stress, 238
  three-dimensional states of, 145
  transformation of axes, 145
  true stress, 109–111, 113, 115–119, 168,
      177, 178, 186–188
  von Mises stress, 206
Stress amplitude, 235–238, 242, 243, 249
Stress concentration factor–fatigue loading
      (K_f), 244
Stress concentration factor–static loading (K_s),
      218, 226, 230
Stress corrosion cracking (SCC), 7, 8, 90, 201
Stress intensity factor, 221–223, 227, 230, 241
Stress–life curves, 263
Stress range, 235, 236, 241–243, 249
Stress ratio, 173, 175, 176, 183, 184, 235, 236,
      242, 243, 249
Stress–strain curves
  for AISI-1020 steel, 46, 56, 57
  elastic, linear-hardening relationship,
      113, 118
  elastic, perfectly plastic relationship,
      113, 118
  elastic, power-hardening relationship, 112
Stress–strain relationships, 3, 45,
      109, 111–112
Substitutional solid-solution alloys, 69, 70, 78
Superalloys, 70, 72, 76, 83, 95, 103, 105, 106,
      258, 264
Surface crack/flaw, 218, 222, 228, 230
Surface defects, 23, 238, 239
Surface residual stresses, 238
Swell ratio (die swell ratio), 130, 140

T
Tangent modulus, 111, 112, 115, 118
Tanker failure, 99, 215
TD-nickel, 72
Tearing mode crack (Mode III), 220
TEM micrograph, 23
Temperature effects in creep, 253, 256–263
Tempering, 85
Tensile strength (ultimate tensile strength), 40,
      44–46, 48, 57, 60, 69, 76, 84, 85,
      88, 90, 92–96, 102, 166, 168, 188,
      202, 208, 212, 236, 238, 243, 244,
      249, 269–271, 274, 277, 279, 280
Tension test
  ductile vs. brittle behavior in, 53
  engineering properties from, 42
  stress-strain curves, 43–46, 56, 57, 60, 110,
      111, 165–166
  test methodology, 47–49
  true stress-strain interpretation of, 109–111

Tertiary stage of creep, 253, 255, 264
Theoretical strength, 22, 36
Theories of failure
    Rankine theory, 201, 202, 208, 213
    Tresca theory, 201–205, 210, 211, 213
    Von-Mises theory, 206–207, 211–213
Theory of linear elasticity, 44, 113
Thermal activation, 253
Thermal strains, 128, 129
Thermal stresses, 23
Thermoplastics, 8, 98–100, 104, 106
Thermosets, 98–100
Three-dimensional states of stress, 153
Three-dimensional stress-strain
        relationships, 153
Time-based growth rate, of cracks, 216
Time-dependent behaviour, *see* Creep
Time-temperature parameters, in creep
    Larson-Miller (L-M) parameter,
        257–258, 263
Time to rupture in creep, 263, 264
Titanic ship failure, 10, 212, 215
Titanium and alloys, 95, 223, 258
Tool steel, 90
Torque, 189–198, 204, 205, 210, 213
Torsional flexibility, 191–192, 196, 198
Torsional stiffness, 192, 196, 198
Torsion of circular shafts, 189, 195, 197
Toughness, from tension tests, 40, 52
Transformation of axes, 145
Transgranular fracture, 7, 9
Tresca criterion, *see* Maximum shear stress
        yield criterion
TRIP steels, 91, 104, 105, 889
True stress and strain, 109
True stress-strain curves, 110
True stress-strain interpretation of tension
        test, 109–111
Tungsten carbide (WC), 48
Tungsten in steel, 48, 89
Twinning, 15, 24, 26–27, 36

U
Ultimate tensile strength, 40, 44, 45, 48, 93,
        166, 168, 188, 202, 208, 238
Uniaxial loading, 166–168, 172–173, 187
Unit cell, 16–22, 30–31, 35, 36

Universal testing machines, 42
Unloading, stress-strain curves for, 121,
        166, 168
Upper yield point (UYP), 165, 166

V
Vacancy, 23, 253
Vanadium in steel, 17, 83, 89
Vickers hardness test, 50, 51, 58, 60
Viscoelasticity, 3, 119–144
Volume fractions, in composites, 271, 274,
        279, 281
Volumetric strain, 124–126, 138, 140,
        141, 144
von Mises stress, 201–207, 211–213

W
Weight percent element, 273
Welding, 39, 64, 215
Welding cracks, 215
White cast iron, 8, 83, 101
Wood, 267
Work hardening, 67, 112, 166, 201, 213
Work hardening exponent *(n)*, 112
Working load, 12, 229
Working stress, 12, 13, 229, 231
Wrought aluminum alloys, 92

Y
Yield criteria, 113–114
Yielding, 29, 45, 113, 121, 165, 166, 168,
        201–203, 205, 206, 210, 213
Yield strength, 4, 8, 22, 29, 34–36, 45, 46, 57,
        59, 66, 68, 72–74, 76, 78, 93, 102,
        105, 121, 131, 141, 166, 168, 194,
        198, 203, 208–210, 212, 213, 253
Young's modulus, 42, 45, 46, 56, 60, 86, 95,
        111, 115, 118, 119, 121, 124, 128,
        142, 217, 218, 224, 226, 230, 231,
        267, 269, 271–275, 277–281

Z
Zinc, 16, 19, 25, 36, 82, 83, 92–94, 105
Zirconia, 96

Printed in the United States
by Baker & Taylor Publisher Services